COLLEEN HARGADEN

ANTHROPOGENIC MINERAL COLLECTION

Special Ef[redacted]fects

ISBN 979-8-9874645-2-6

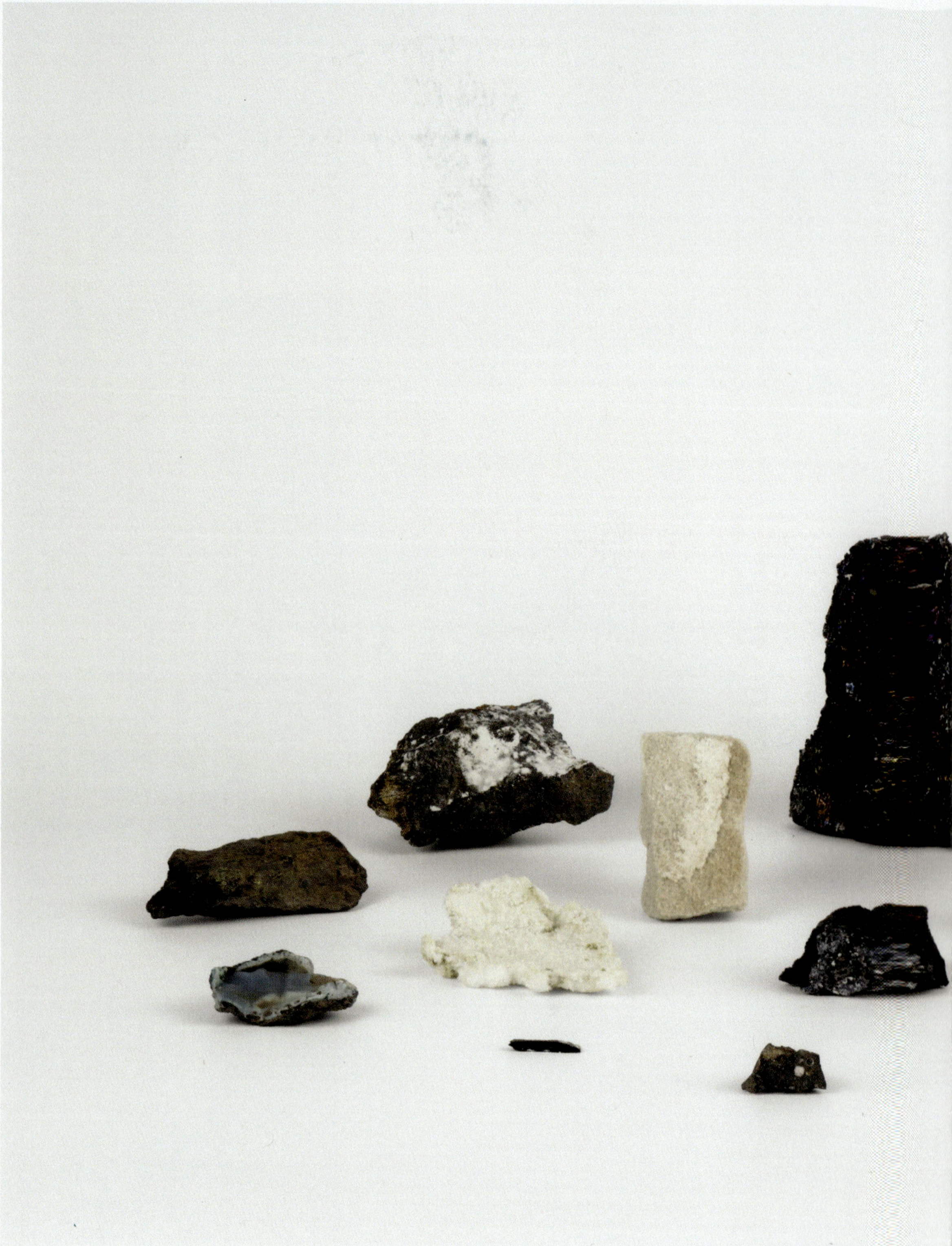

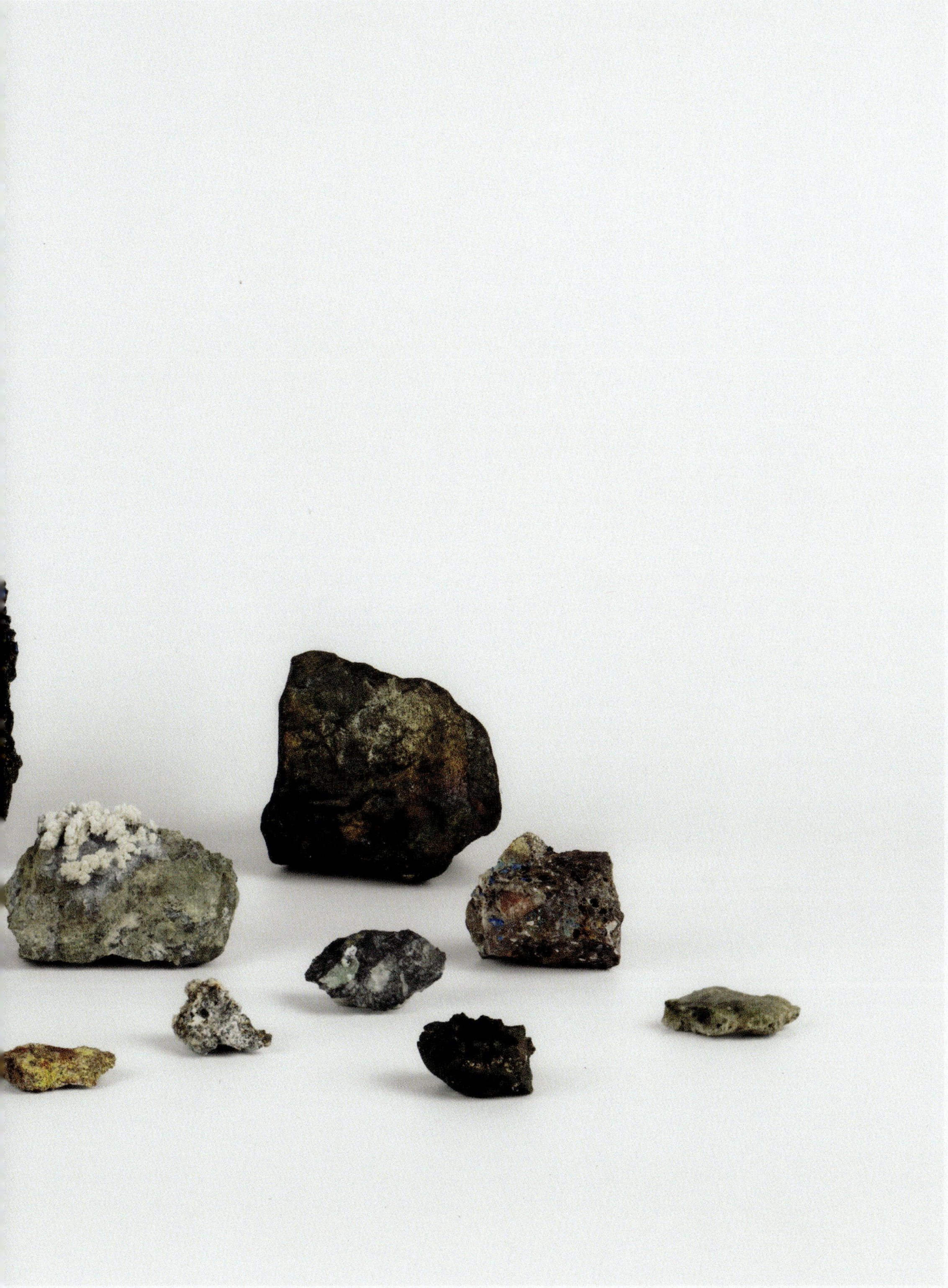

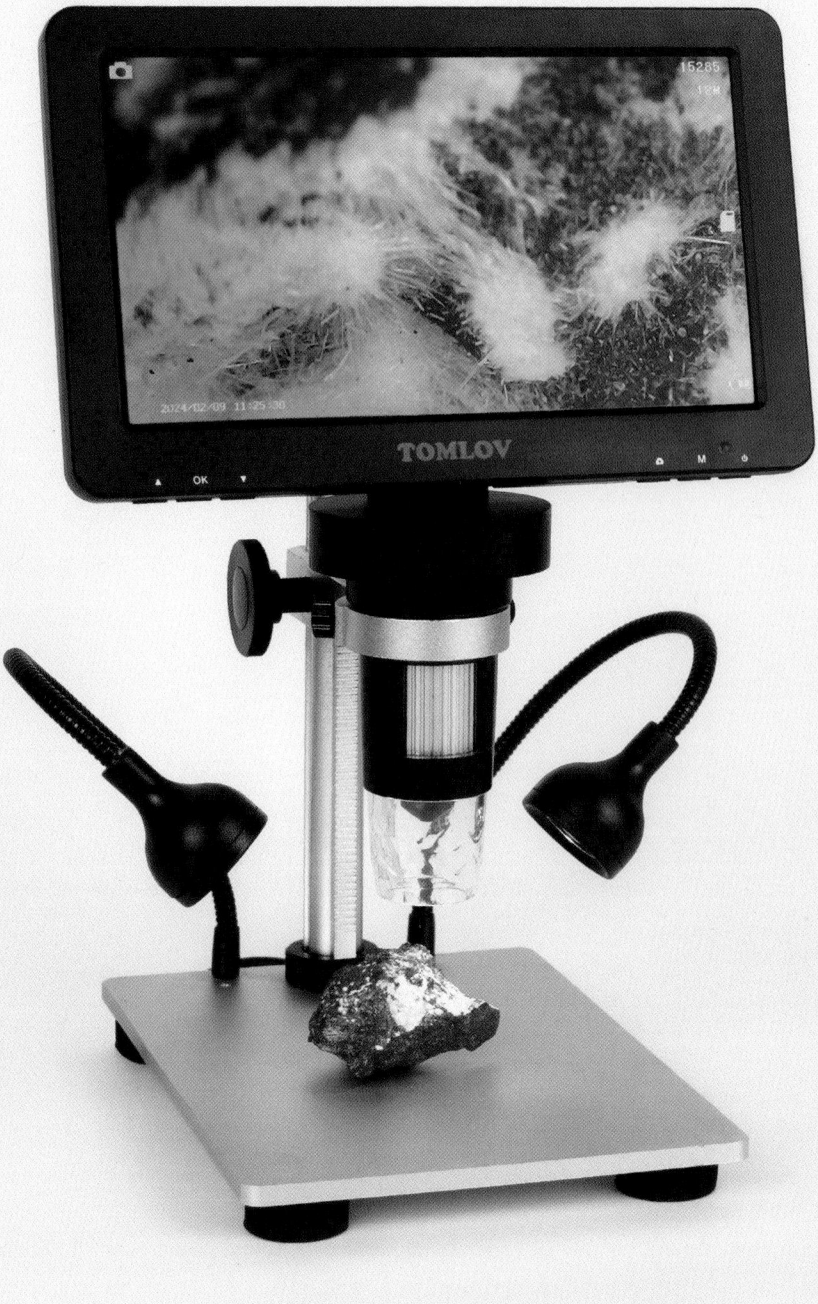
15285
2024/02/09 11:25:38
TOMLOV
OK
M

Introduction by Colleen Hargaden

ANTHROPOGENIC MINERAL COLLECTION

In the fall of 2020, I began gathering samples for what I have termed an “anthropogenic mineral collection.” These minerals are human byproducts caused by our geological intervention spanning from the Industrial Revolution to the present day. The majority of minerals in my collection occur inadvertently at mining sites where valuable metals are extracted during human excavation. This process can be initiated when geological material is displaced during tunnel construction and undergoes oxidation, weathering, or combustion over time. These new minerals will be preserved in the geological record as distinctive of the human age, marking our desire to harvest the Earth’s energy and extract its materials.

Many of the minerals in my collection were found in ore dumps and furnace slag, along the tunnel walls of mines and smelters, and within the debris of human-induced fires. Their localities span the United States, Poland, Italy, Germany, England, Austria, and Greece, and new minerals continue to be discovered internationally. Despite the small size of these mineral samples, their existence reveals an often obscured global system of production. The industrial sites where anthropogenic minerals are found extract metals used in a wide range of technologies. For instance, the very apparatuses I use to document these samples—my digital microscope, camera, and laptop—all utilize metals sourced from mines similar to those where my collection’s minerals originate.

Each sample has been gathered second-hand from amateur geologists, mineralogists, abandoned mine foragers, and rockhounders across the globe. These individuals sell their findings on websites like Ebay or E-Rocks.com, with the average cost of each sample ranging between $20-30. Certain sellers will repeatedly list anthropogenic minerals, and over time, I have developed a back and forth with them. I am a repeat customer of New Brand Minerals, based out of Italy, which is operated by Marzio Mamberti. Mamberti was particularly helpful when it came to addressing my safety concerns regarding “hot rocks,” or radioactive samples, of which I have a few (Bayleyite, Andersonite, and Trinitite). I was nervous when I began to accumulate multiple hot rocks, but he claimed that even if I were “to keep 100 samples on my desk and expose myself to them 10 hours a day, everyday, it would be unlikely to reach the amount of radiation caused by a normal x-ray.”

Anthropogenic Mineral Collection

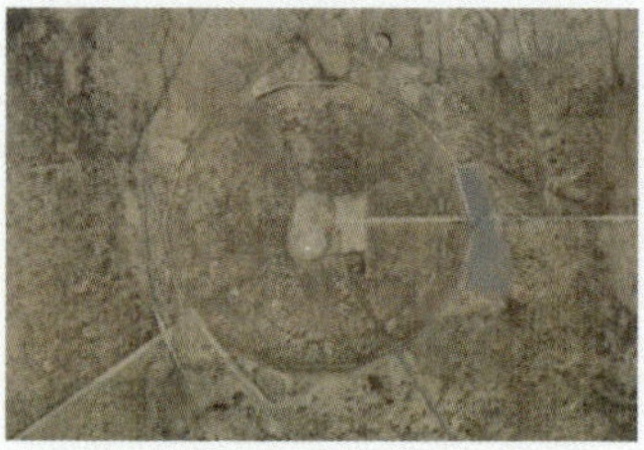

A

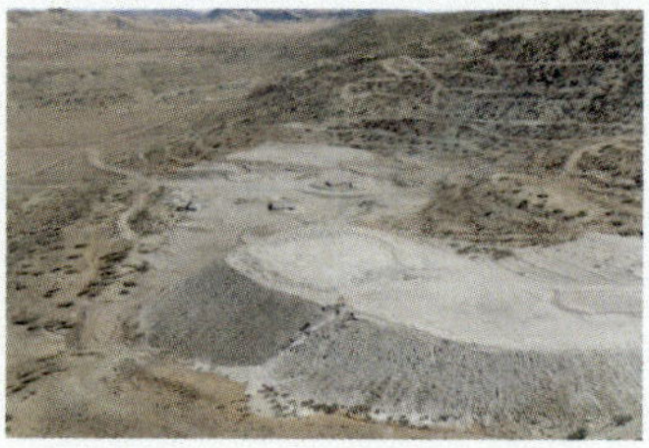

B

My anthropogenic mineral collection is guided by an article published in March 2017 in American Mineralogist by Robert Hazen, Edward Grew, Marcus Origlieri, and Robert Downs.[1] The study catalogs the 208 new mineral species that occur principally or exclusively as a consequence of human processes, and notes that even more species are forming at an unparalleled rate. Robert Downs, professor of Mineralogy at the University of Arizona, states that,

> Given humanity's pervasive influences on the environment, there must be hundreds of as yet unrecognized "minerals" in old mines, smelters, abandoned buildings, and other sites. Meanwhile, new suites of compounds may now be forming in, for example, solid waste dumps where old batteries, electronics, appliances, and other high-tech discards are exposed to weathering and alteration.[2]

The "e-waste" which Downs alludes to continues to accumulate as recycling companies along the American Pacific Coast opt to ship piles of discarded electronics overseas. These products are manufactured in Asia and shipped to the states via the West Coast, they are consumed by Americans, and are returned to Asia in the form of e-waste. This industrial loop is facilitated by the Asia-Pacific Economic Cooperation (APEC) which produces approximately eighty percent of the world's computers and high tech components, as well as its toxic refuse.[3] The mineral and chemical interactions that occur within e-waste dumps have yet to be extensively studied.

The line between an anthropogenic mineral and non-mineral is blurry. Human intervention in the creation of a mineral is often indirect, for instance, when it results secondarily from human activities like excavation. Others occur directly through interactions between geological material and human created material, such as melted plastic or corroded metal. While the International Mineralogical Society (IMS) does recognize anthropogenic minerals as a species, they generally do not regard substances created by combustion, or fires, as minerals.[4] Geological samples that occur through combustions at coal mines, waste dumps, and other industrial sites are contentiously debated by the IMS. In my collection, I nonetheless include "minerals" that result from human-induced fires such as mine fires, campfires, and most notably, the Trinity nuclear test.

[1] Robert M. Hazen, Edward S. Grew, Marcus J. Origlieri and Robert T. Downs
On the mineralogy of the "Anthropocene Epoch" (American Mineralogist, March 2017)

[2] *Catalog of 208 Human Caused Minerals Bolsters Argument for 'Anthropocene Epoch'* (Carnegie Science, March 2017)

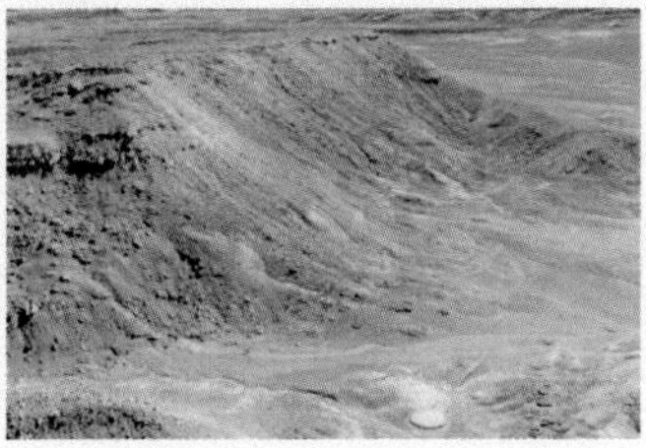

C

The challenge of categorization is not new to geologists. For example, the proposal of an "Anthropocene Epoch" is contested among many scientists, primarily due to the lack of agreement on when precisely humans began impacting geological stratification.[5] Some believe it was the Trinity nuclear bomb test of 1945, others name the Industrial Revolution, or the advent of agriculture. The geological record is being impacted by humans, but unequivocally naming the beginning of the epoch constrains the conversation. Anthropogenic minerals resist definition and push the boundaries of the geological field as a whole, sparking questions regarding different levels of human responsibility in their origin.

The seventeen mineral profiles in this guide feature descriptions, microscopic and macro photography, location coordinates, and 3D renderings based on aerial imaging of the site each mineral was found. This anthropogenic mineral collection currently consists of Andersonite, Bayleyite, Canavesite, Dypingite, Fiedlerite, Hydromagnesite, Iron Ore Slag, Lansfordite, Metamunirite, Nesquehonite, Plastiglomerate, Phosgenite, Rainbow Slag, Shannonite, Simonkolleite, Tinnunculite, and Trinitite. A full list of the 208 anthropogenic mineral species, as published by American Mineralogy in 2017, can be located at the back of this guide.

A Trinity Site, New Mexico
B Sunday Mine, Colorado
C Yellow Cat District, Utah

[3] Rosenberg, Matt. *Pacific Rim and Economic Tigers* (ThoughtCo. January, 2020)

[4] Nickel, E.H, Grice, J.D, IMA Statement on *"anthropogenic substances"* (Canadian Mineralogist, 1998)

[5] Zhong, Raymond. *Are we in the "Anthropocene," the Human Age? Nope, Scientists Say.* (New York Times, 2024)

INDEX

ANDERSONITE

DESCRIPTION:

This piece of Andersonite comes from the tunnels and timbers of the Atomic King Mine in Moab, Utah, United States. The Atomic King Mine extracted uranium ore, and Andersonite became the byproduct of mixing sandstone, uranium, and moisture in the mine tunnels. Andersonite is known for its bright green fluorescent glow under a black light which is evidence of its radioactivity.

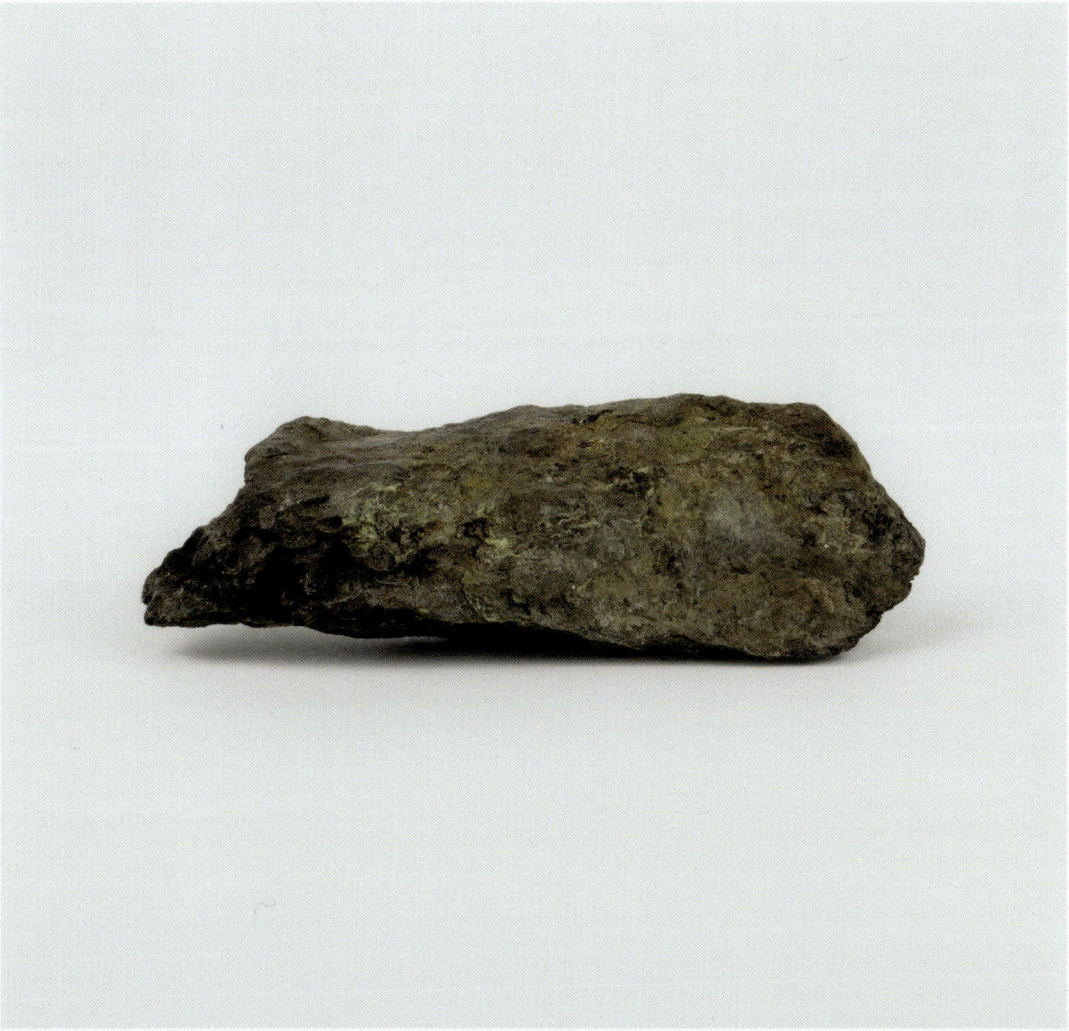

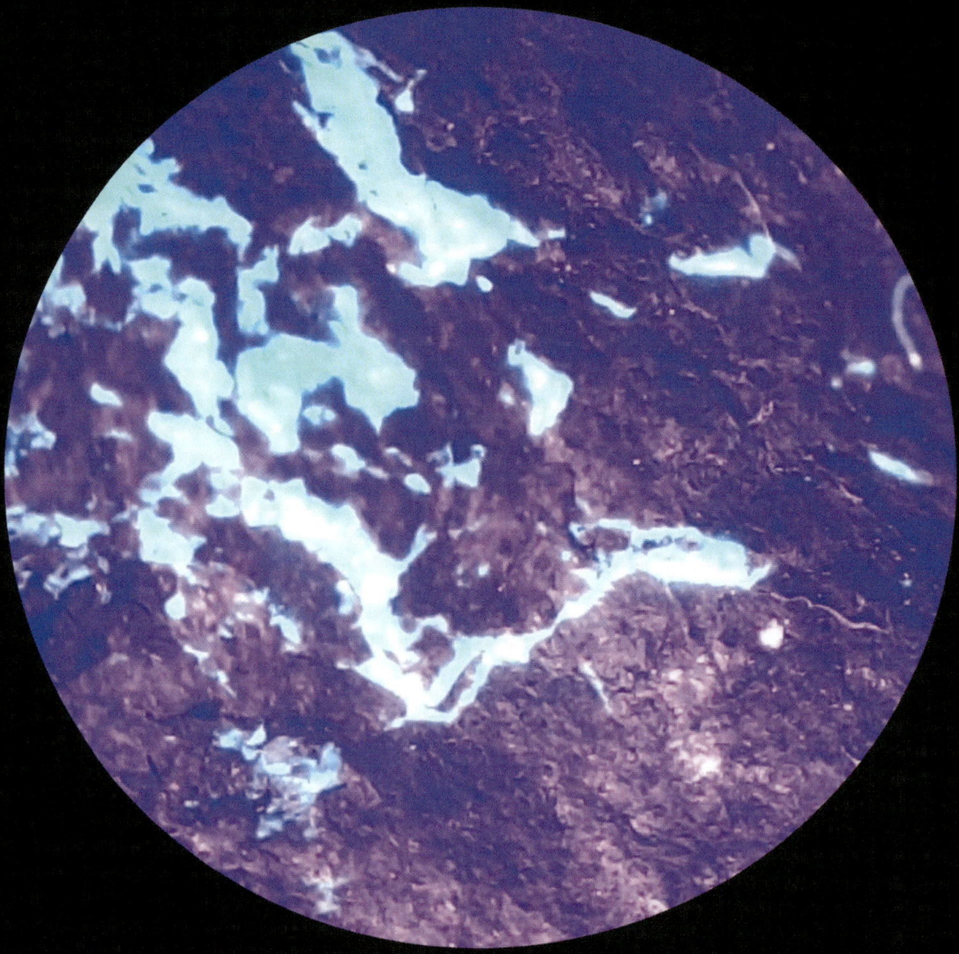

$Na_2Ca(UO_2)(CO_3)_3 \cdot 6H_2O$

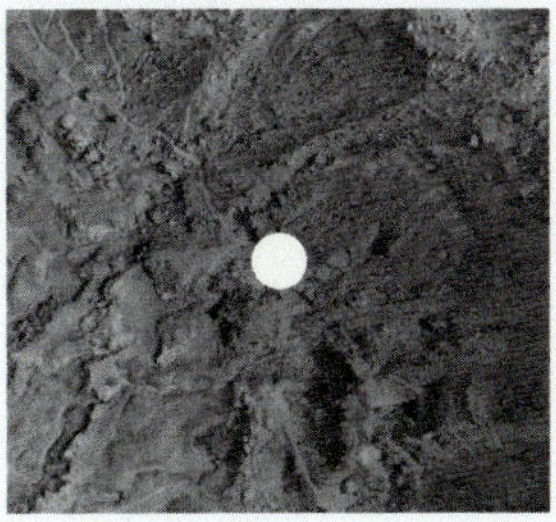

LOCALITY: Atomic King Mine, Moab, San Juan County, Utah, United States

COORDINATES 38.47804, -109.59737

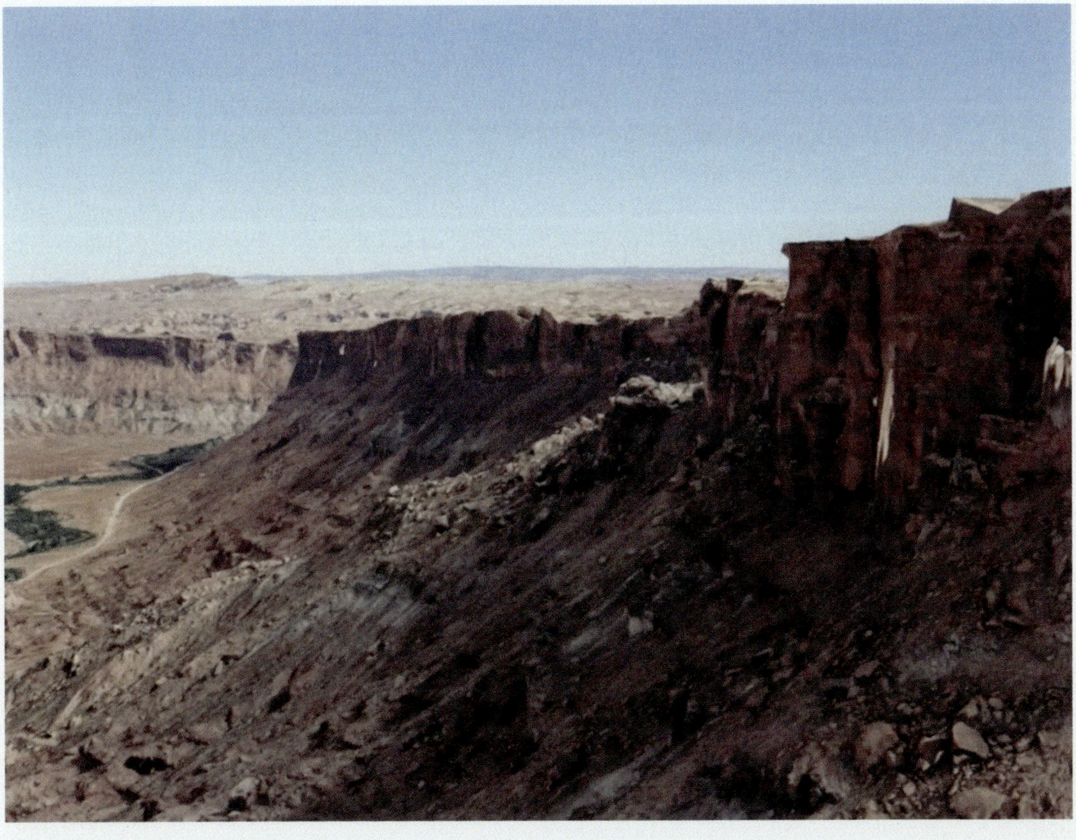

BAYLEYITE

DESCRIPTION:

Bayleyite was first observed and collected from the oxidation zone of a lead and gold mine known as the Hillside Mine in Arizona, United States. It is a secondary mineral which contains magnesium, uranium and carbon. Bayleyite has a bright yellow color, and is found in the form of crust on uranium bearing ores.

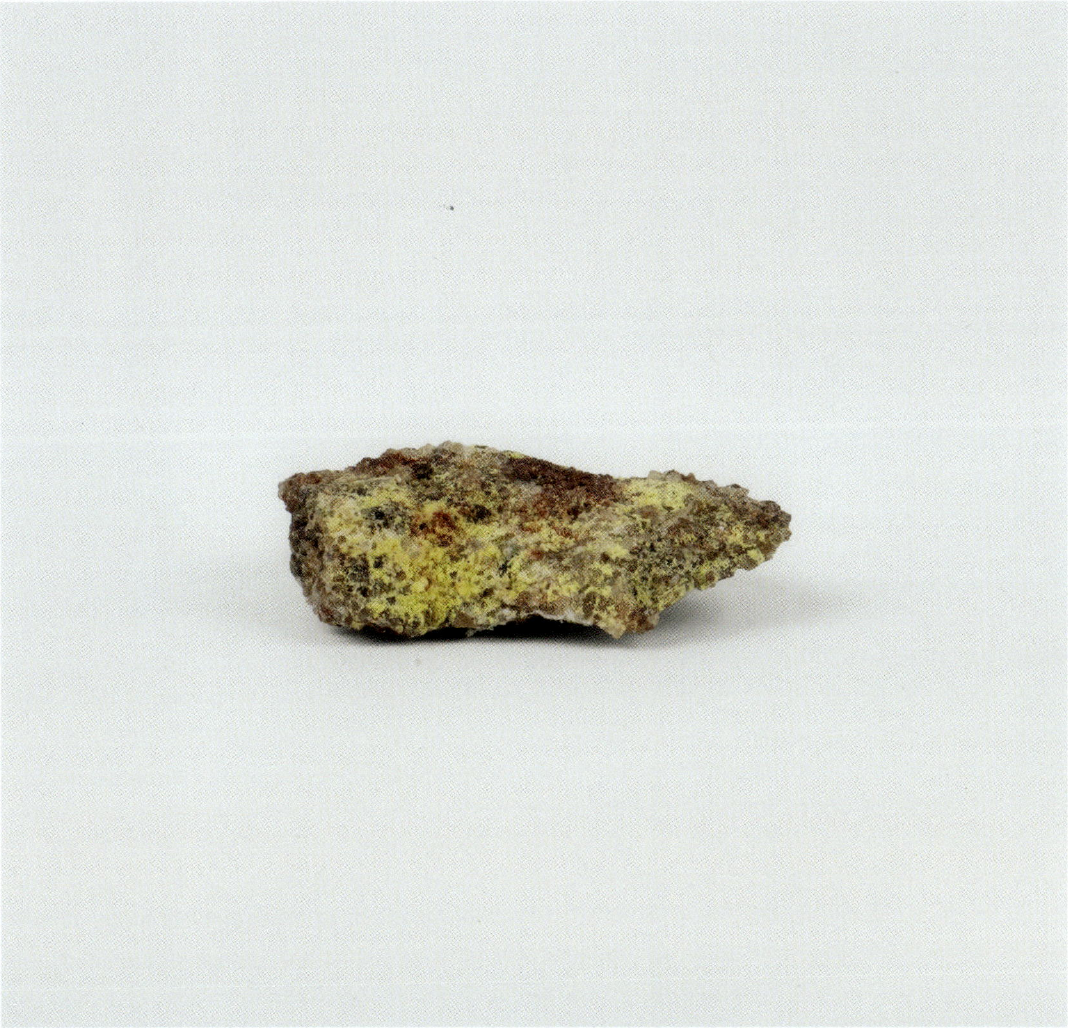

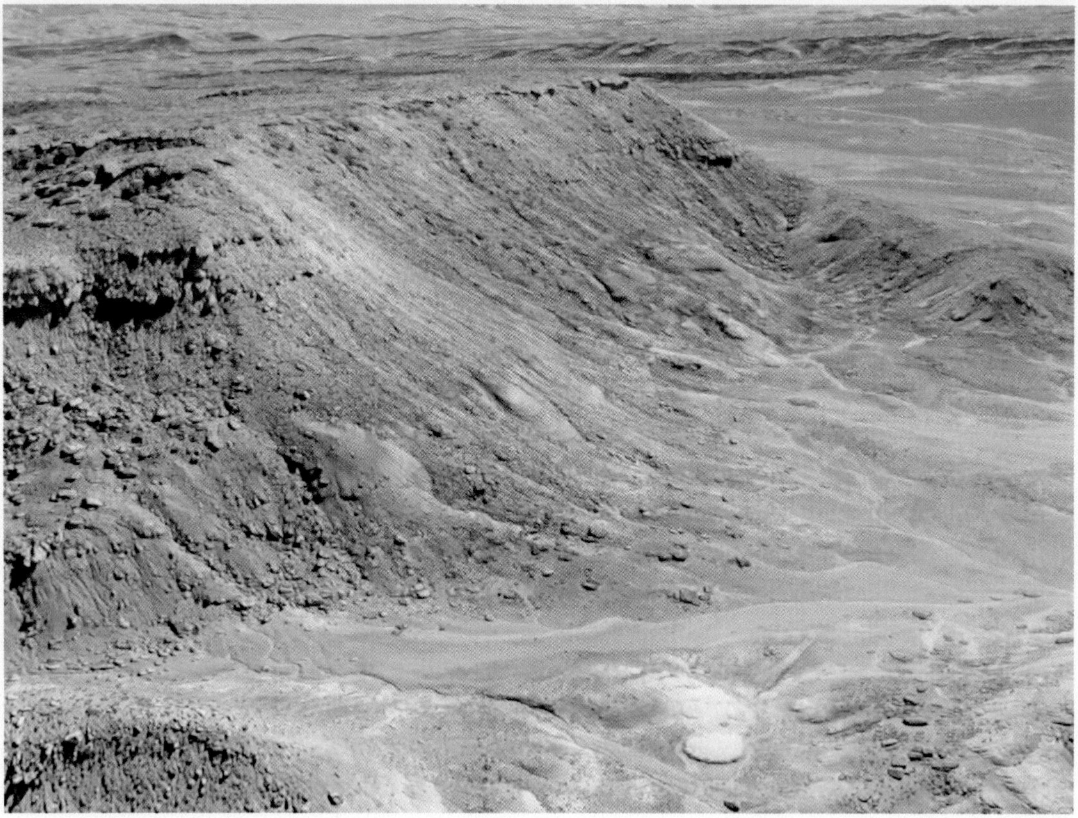

LOCALITY: Yellow Cat District, Grand County, Utah, United States

COORDINATES 38.86276, -109.45459

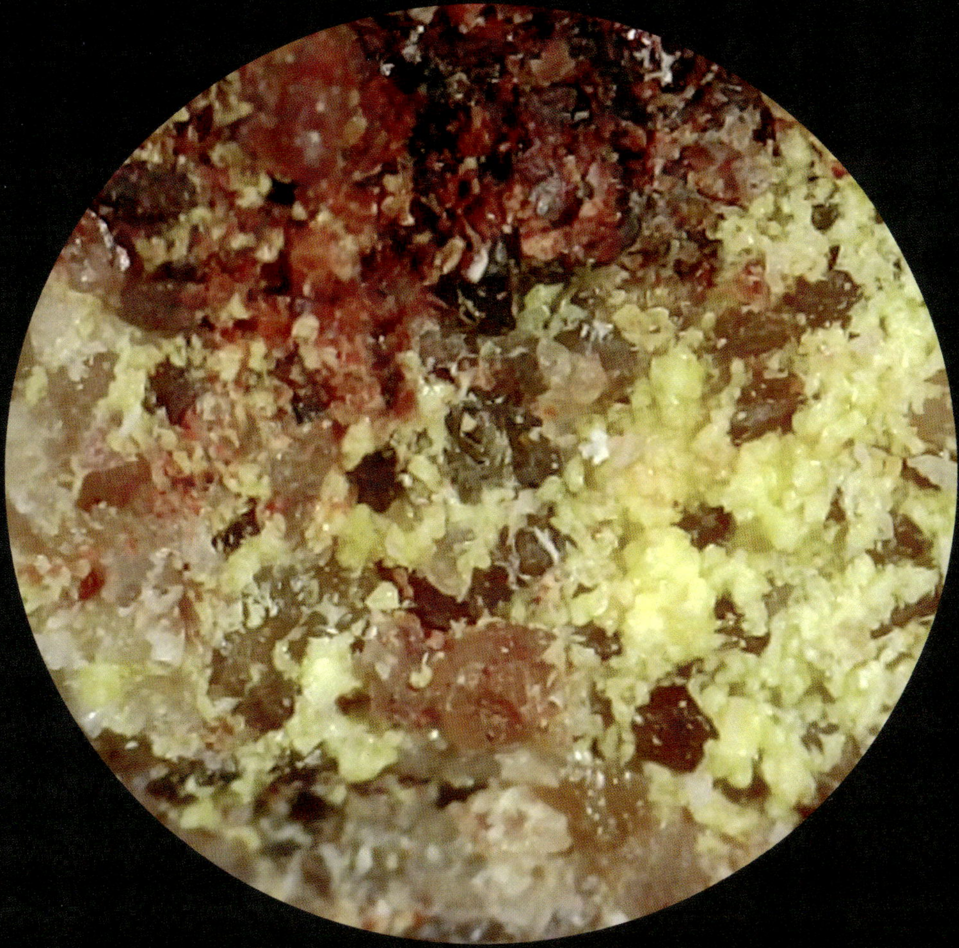

$Mg_2(UO_2)(CO_3)_3 \cdot 18H_2O$

CANAVESITE

DESCRIPTION: This piece of Canavesite is from an abandoned pyrite mine in Brosso, Italy - just outside of Torino. Canavesite is a human-byproduct, created by the weathering of ludwigite-magnetite skarn on the walls of pyrite mines. The only other place on earth Canavesite has been found is in an iron extraction site known as the Sterling Mine in New Jersey, United States.

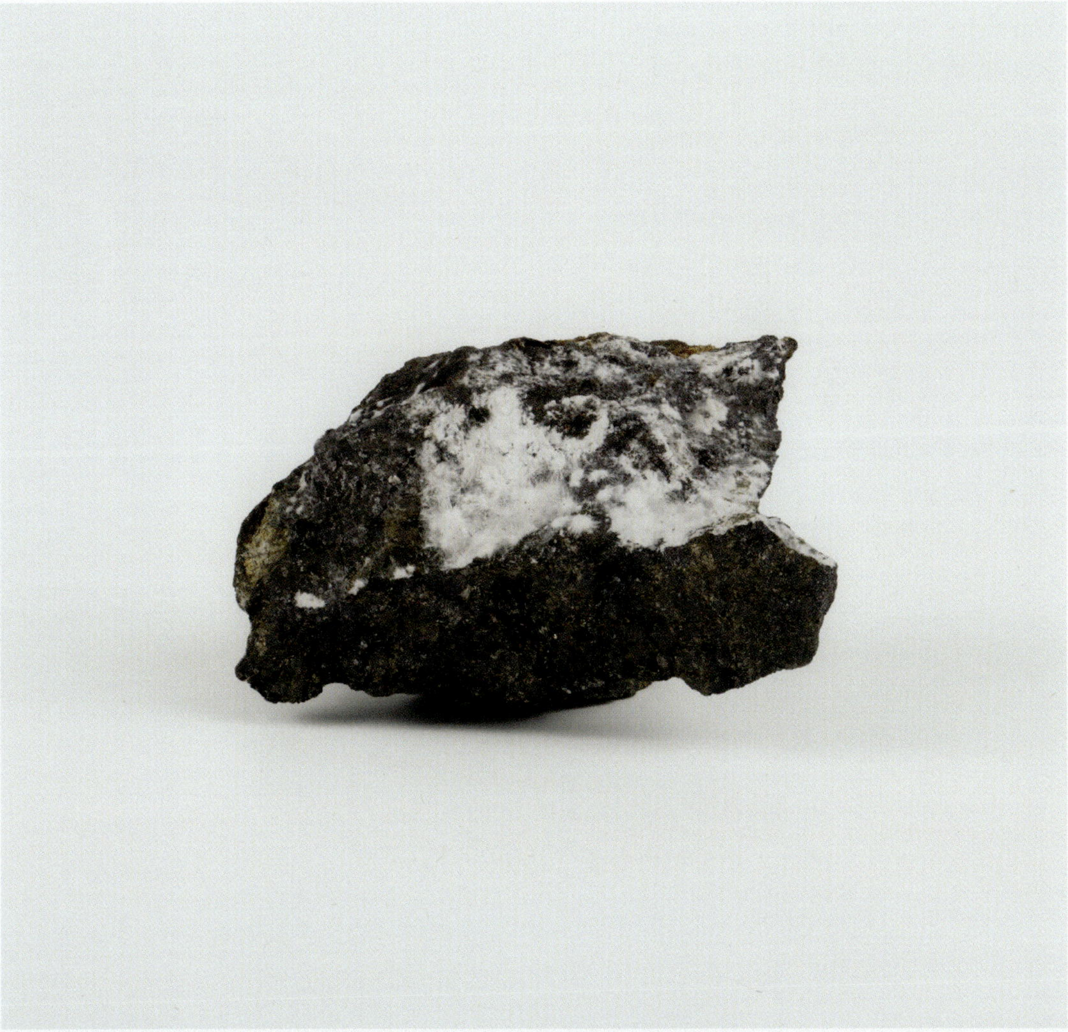

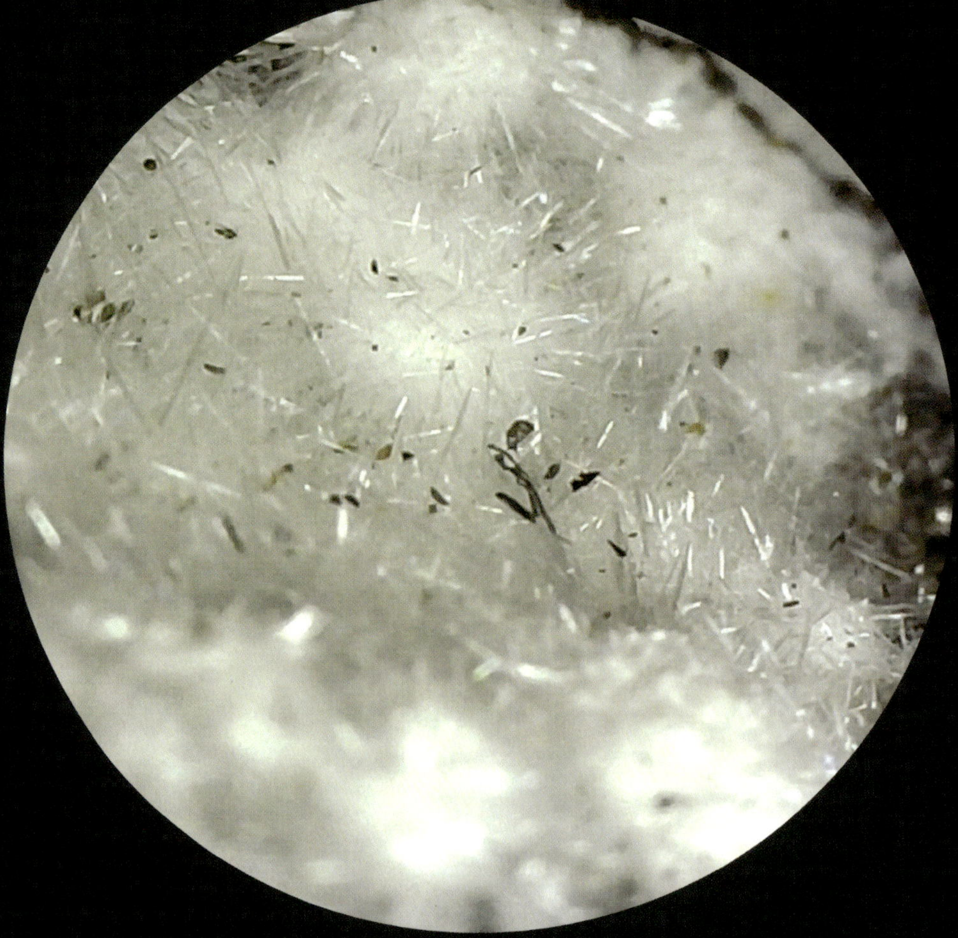

$Mg_2(HBO_3)(CO_3) \cdot 5H_2O$

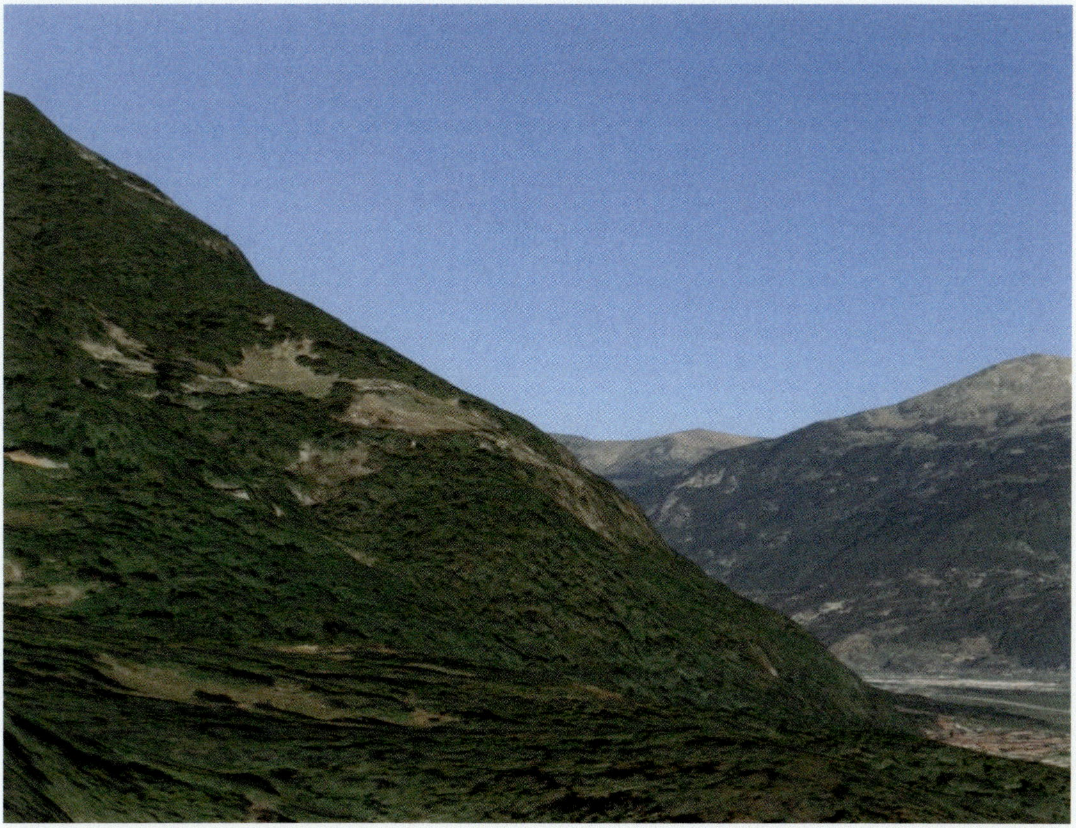

LOCALITY: Brosso Mine, Calea, Lessolo, Piedmont, Italy

COORDINATES 45.49946, 7.81750

DYPINGITE

DESCRIPTION: The slag heap of the Marcel coal mine in Radlin, Poland, contains the spoils that are being thrown away during the coal excavation. The waste and active chemical substances contained in materials stored at the site have caused fires, as well as toxic smoke and gas in the area. Dypingite is a mineral byproduct created during these fires, which sublimates from escaping gas.

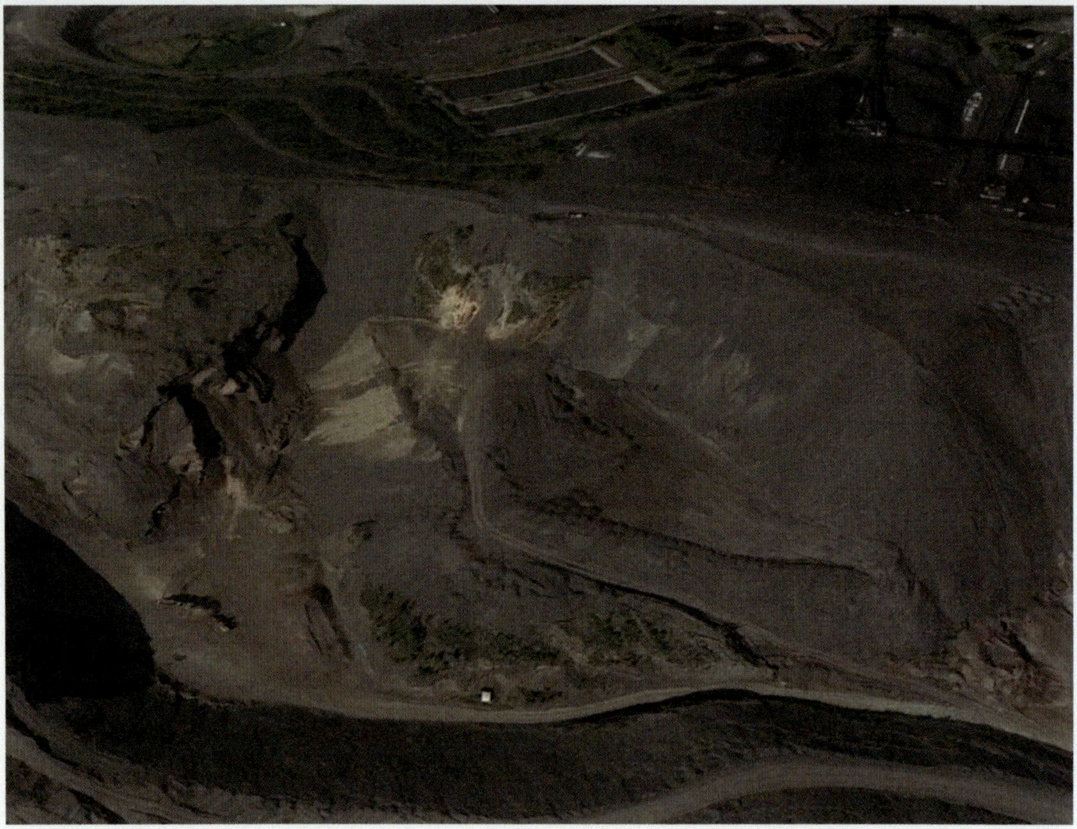

LOCALITY: Marcel Coal Mine, Radlin, Poland

COORDINATES 50.041667, 18.486944

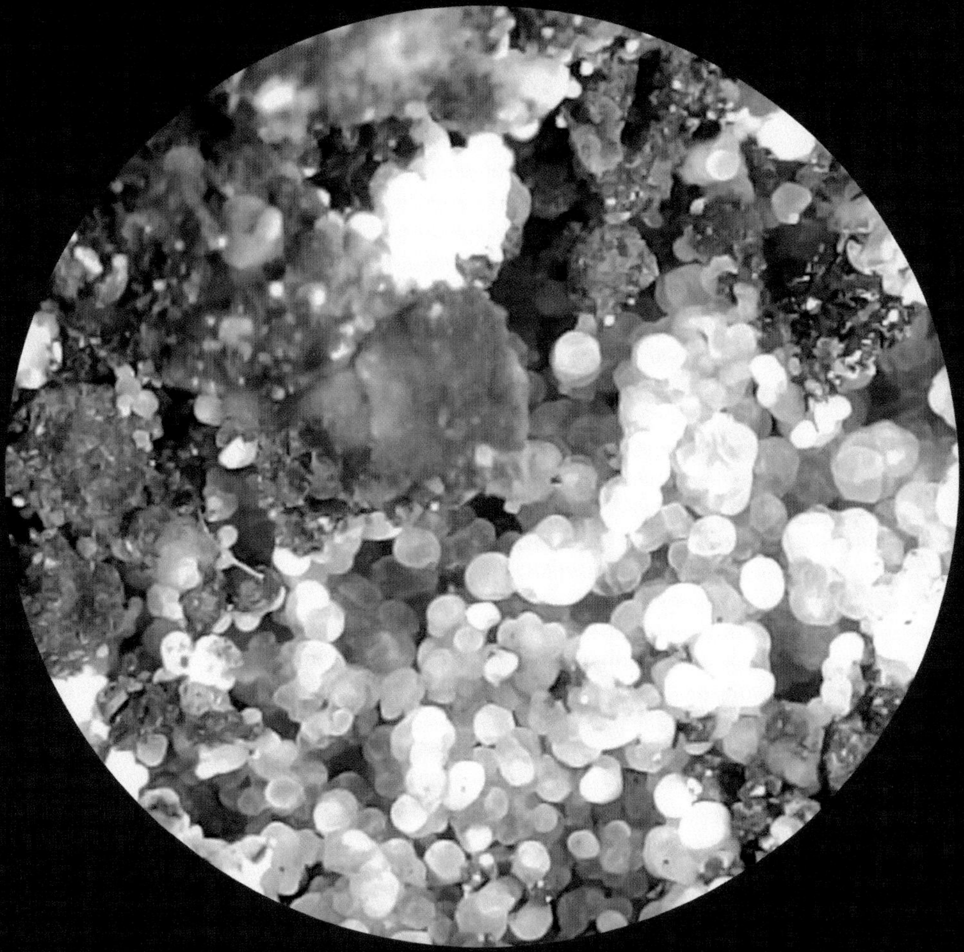

$Mg_5(CO_3)_4(OH)_2 \cdot 5H_2O$

FIEDLERITE

005

DESCRIPTION:

Fiedlerite was first found in the ancient slag dumps of Lavrio, Greece. Leftover rocks from lead mines, too poor in metal content to be processed, were dumped into the sea. Over time, the sea water altered these low grade lead ores and produced an assortment of rare minerals, including Fiedlerite. Contemporary technology now has the ability to extract valuable metal from these salt-saturated ores, which has led to the reprocessing of old dumps and flooded mines in Lavrio. This sample was found in a lead mine known as the Gannel Smelter in Cornwall, England.

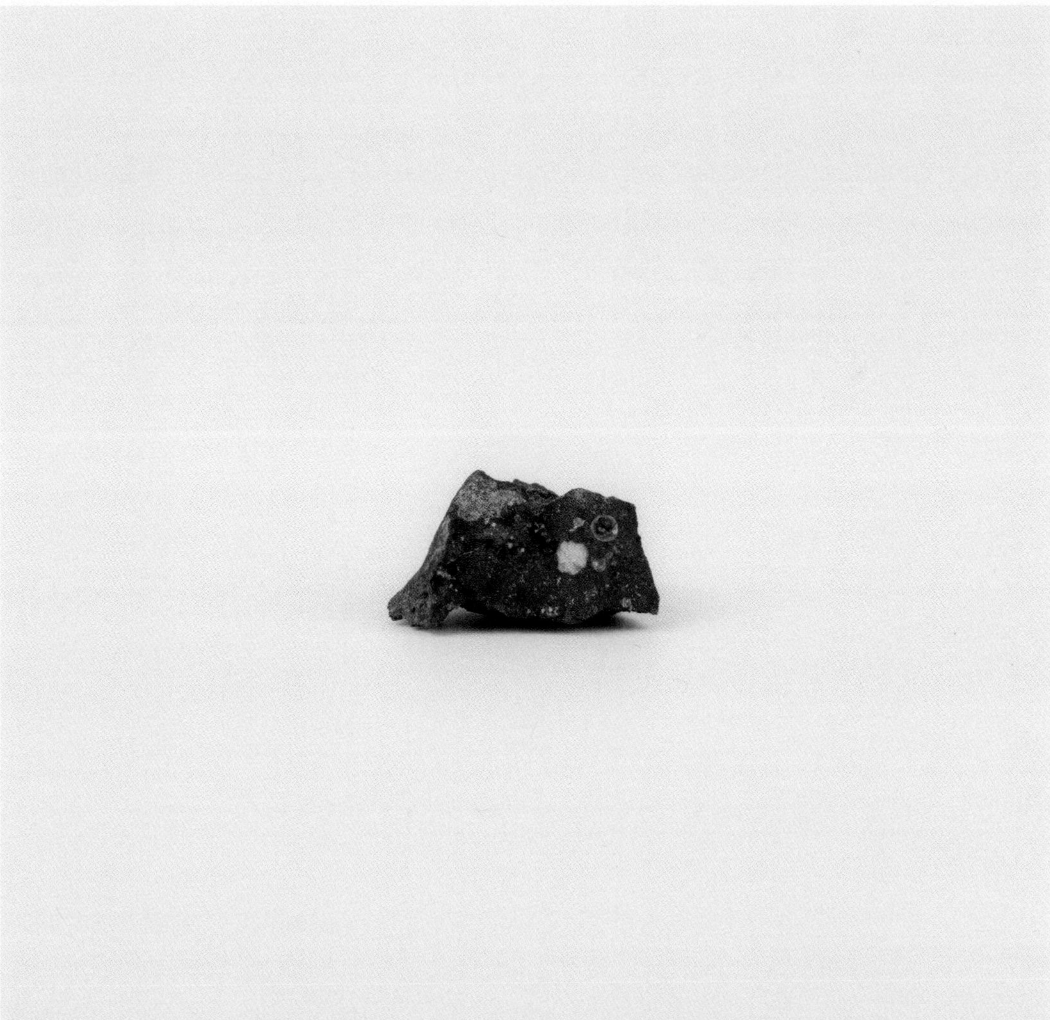

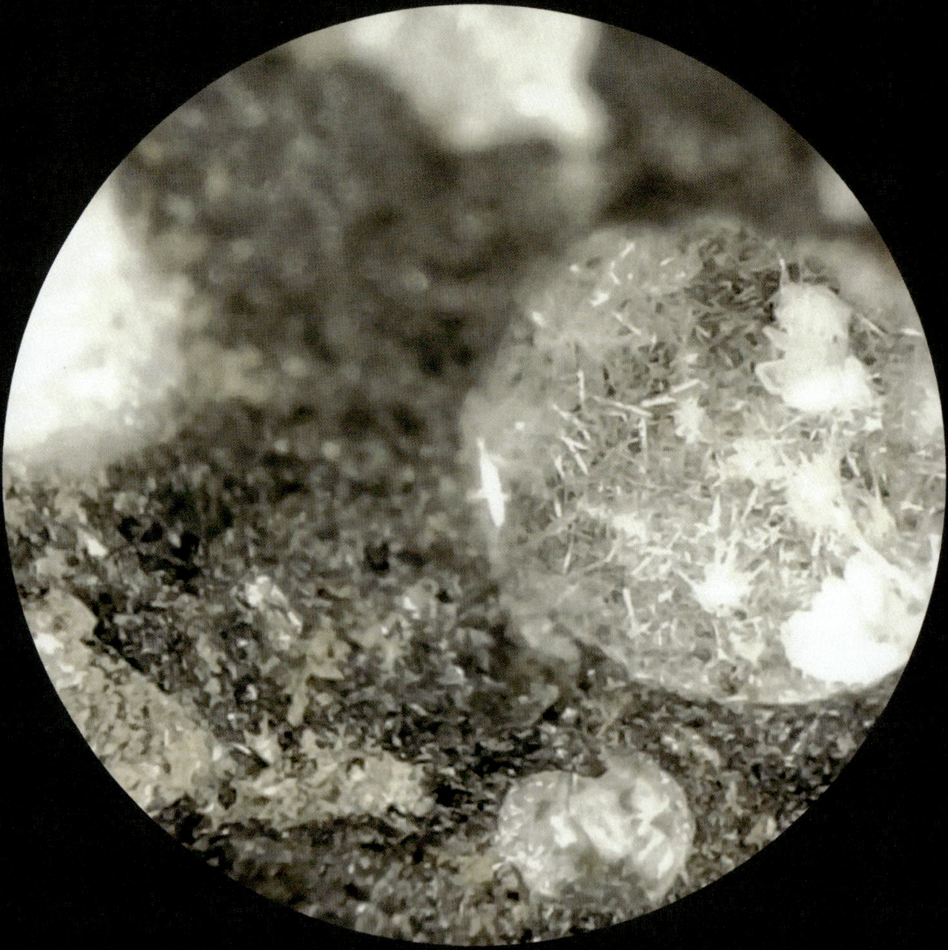

$Pb_3FCl_4(OH) \cdot H_2O$

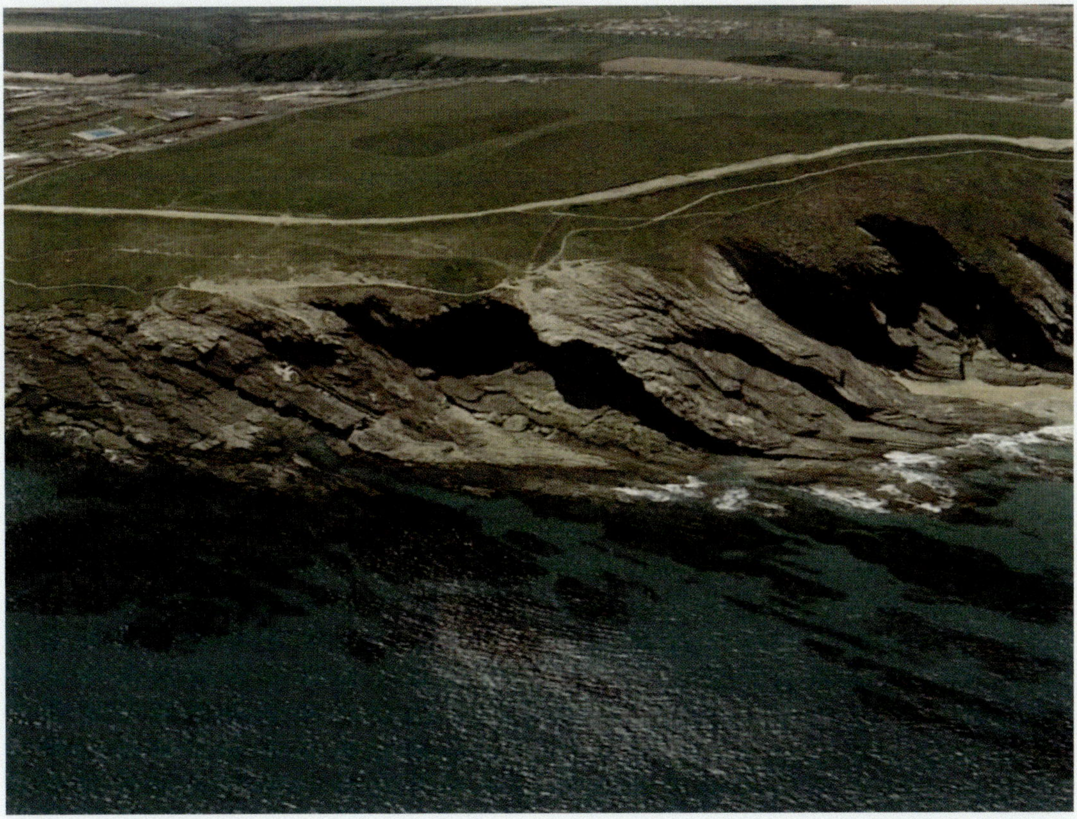

LOCALITY:	Gannel Smelter Mines, Crantock, Cornwall, England
COORDINATES	50.40847, -5.10931

HYDROMAGNESITE

DESCRIPTION:

Hydromagnesite forms out of a substance sometimes found in caves which never hardens called “moonmilk.” This sample is recovered from a Graphite Mine dump in Austria. Hydromagnesite is sometimes collected and used as a fire retardant, while graphite from the mine is commonly used in solar panels and batteries.

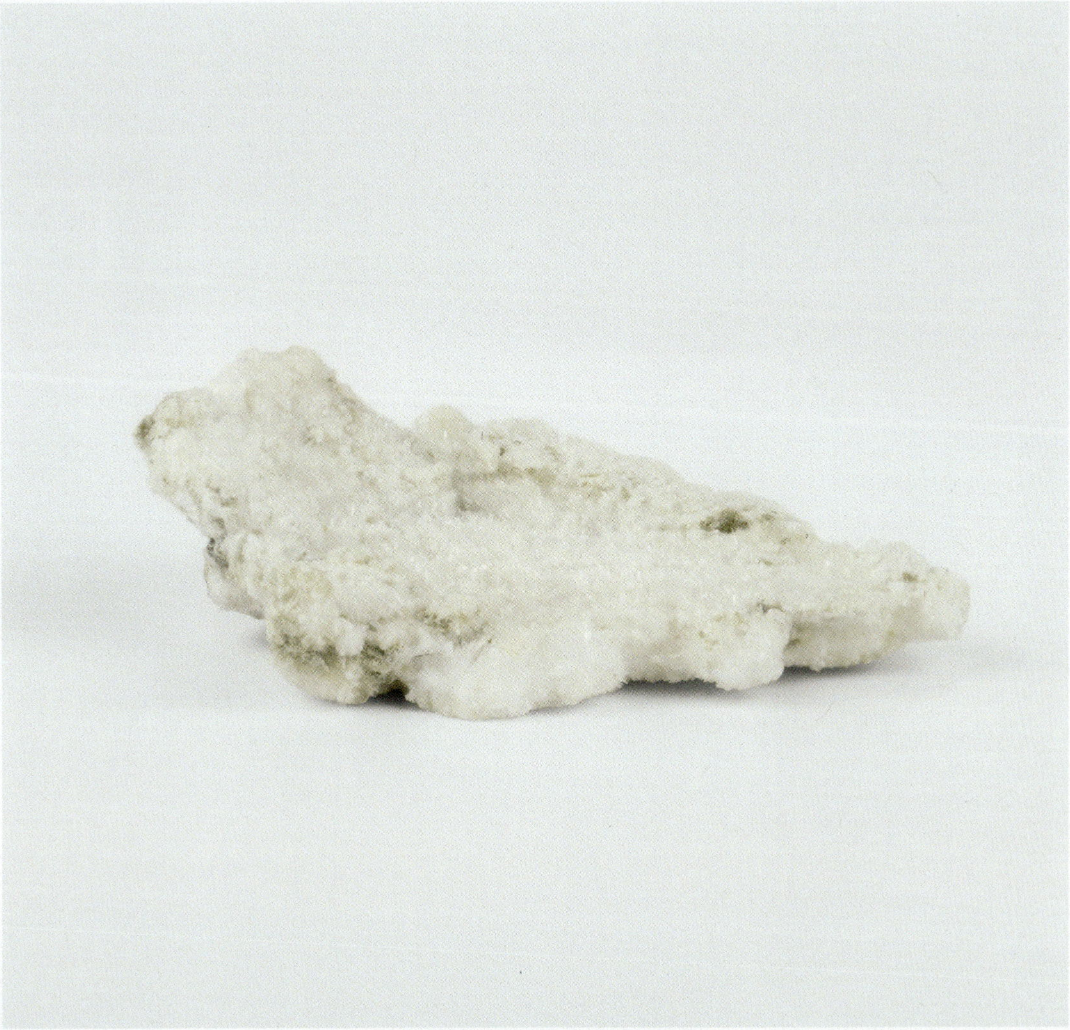

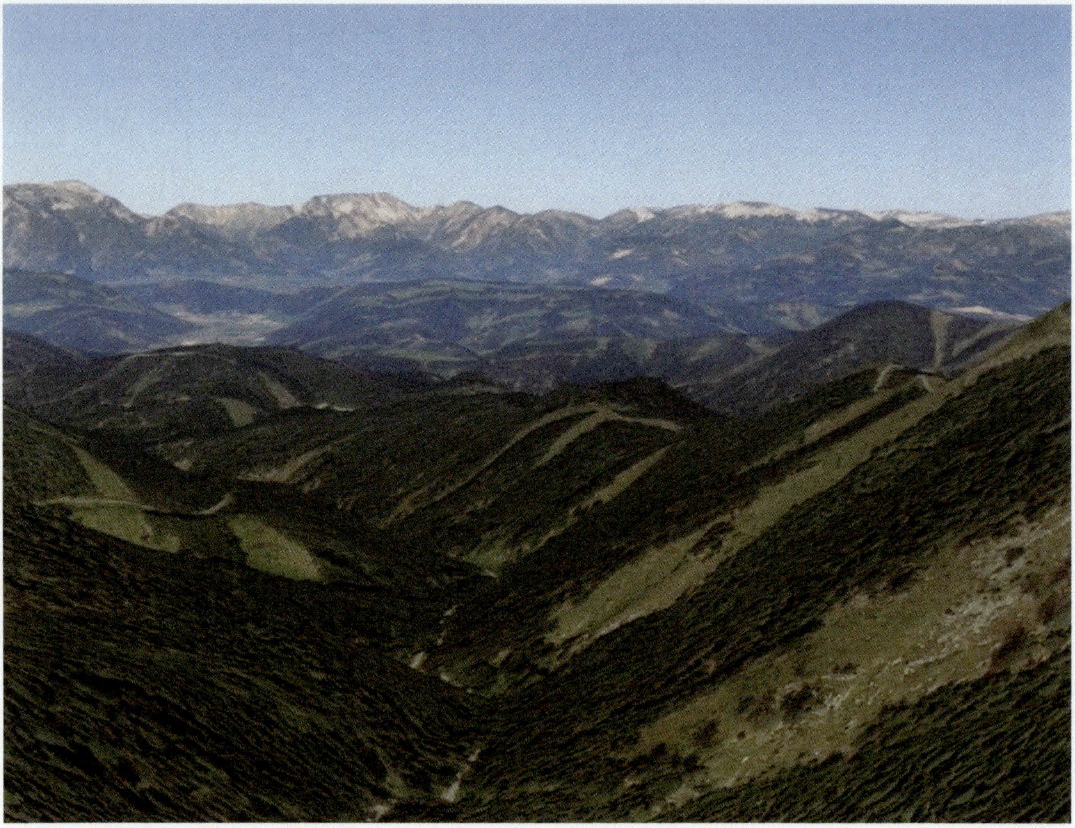

LOCALITY: Lobminggrabben, St Stefan, Styria, Austria

COORDINATES 47.27365, 15.07525

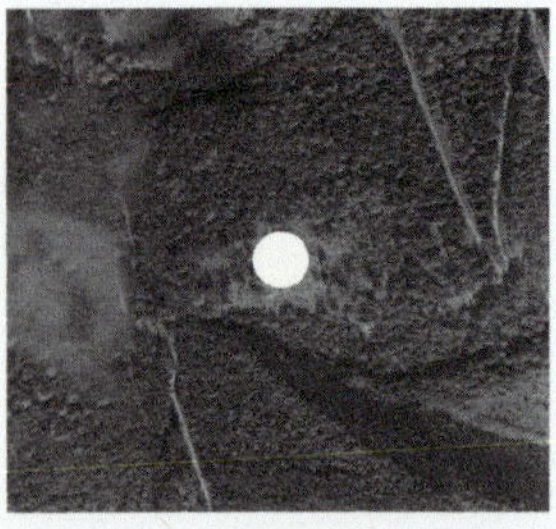

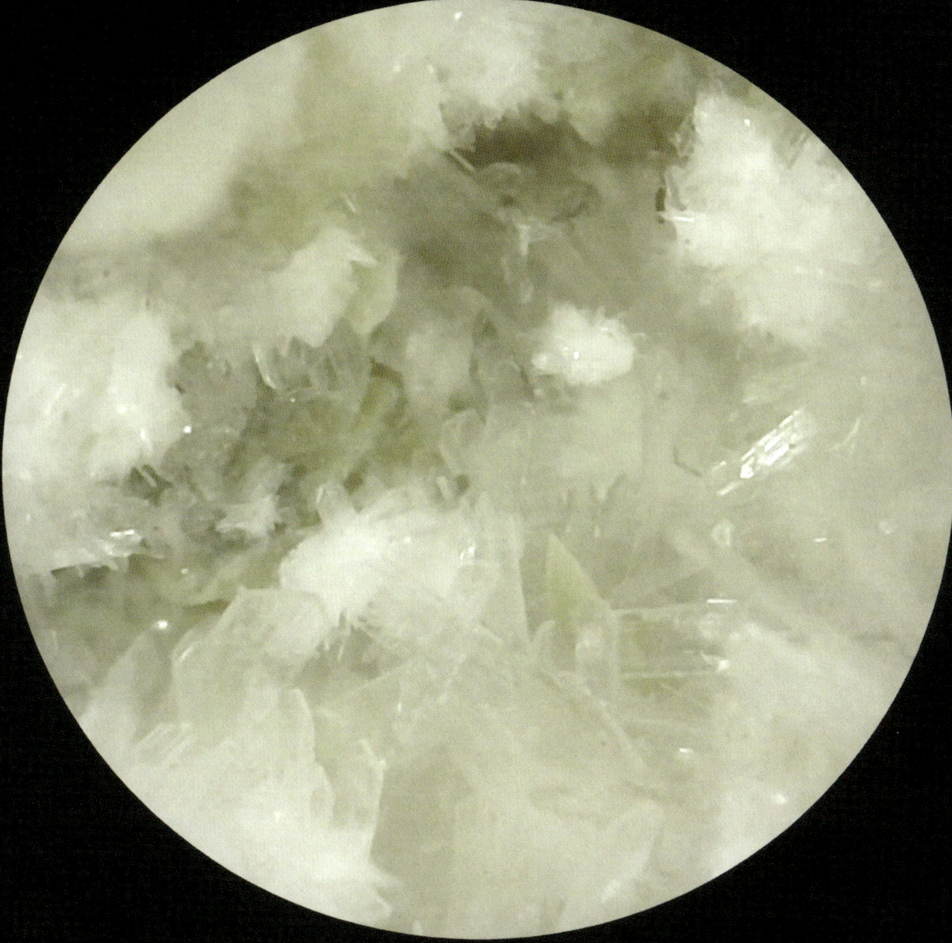

$Mg_5(CO_3)_4(OH)_2 \cdot 4H_2O$

IRON ORE SLAG

DESCRIPTION:

This air-cooled Iron Ore Slag is a coproduct of iron manufacturing on Zug Island in Michigan, United States. During production, the blast furnace is charged with iron ore, fluxing agents (usually limestone and dolomite) and coke as a fuel and reducing agent. The coke combusts to produce carbon monoxide, and this process reduces the iron ore to a nonmetallic slag.

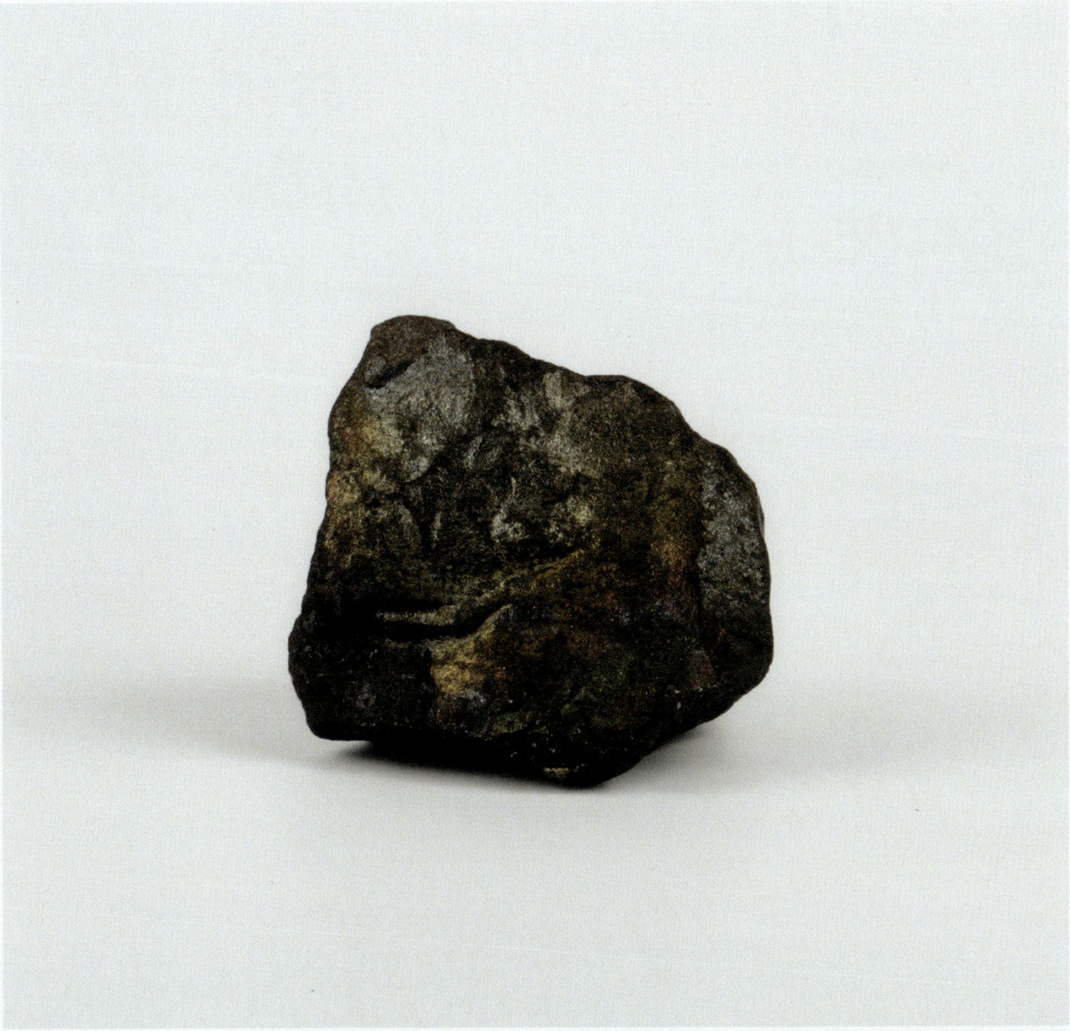

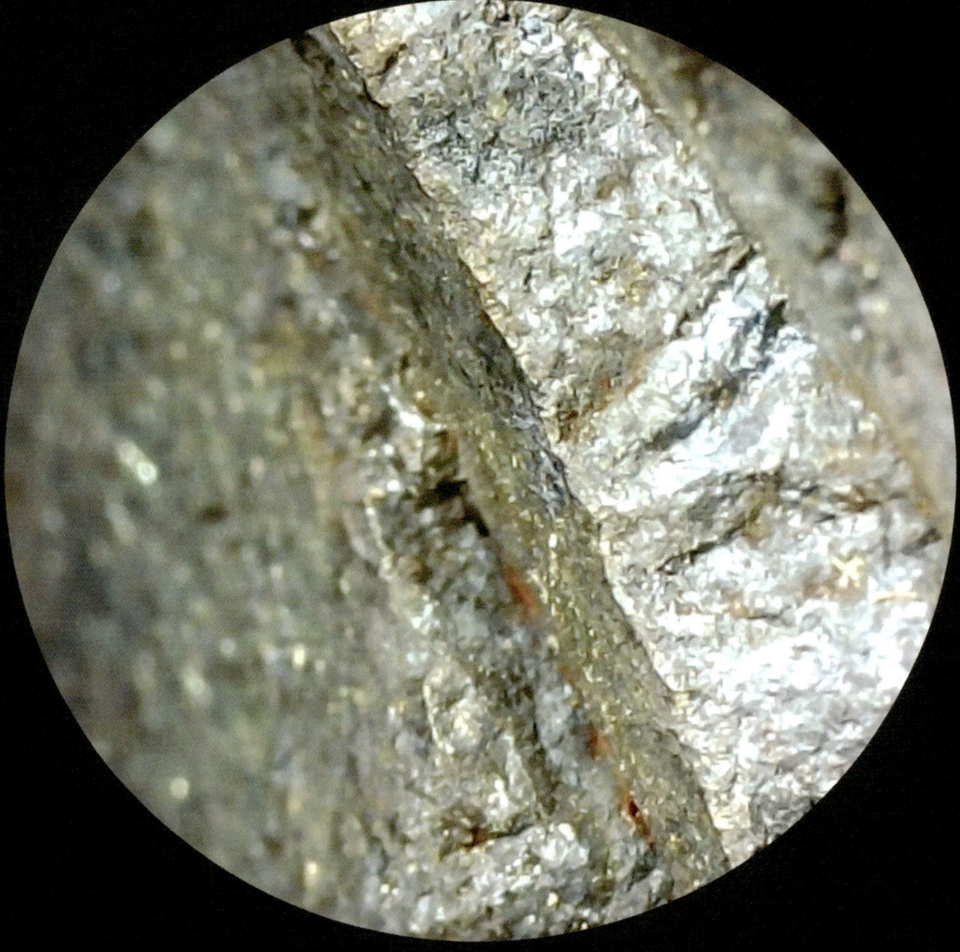

$CaO \cdot SiO_2$

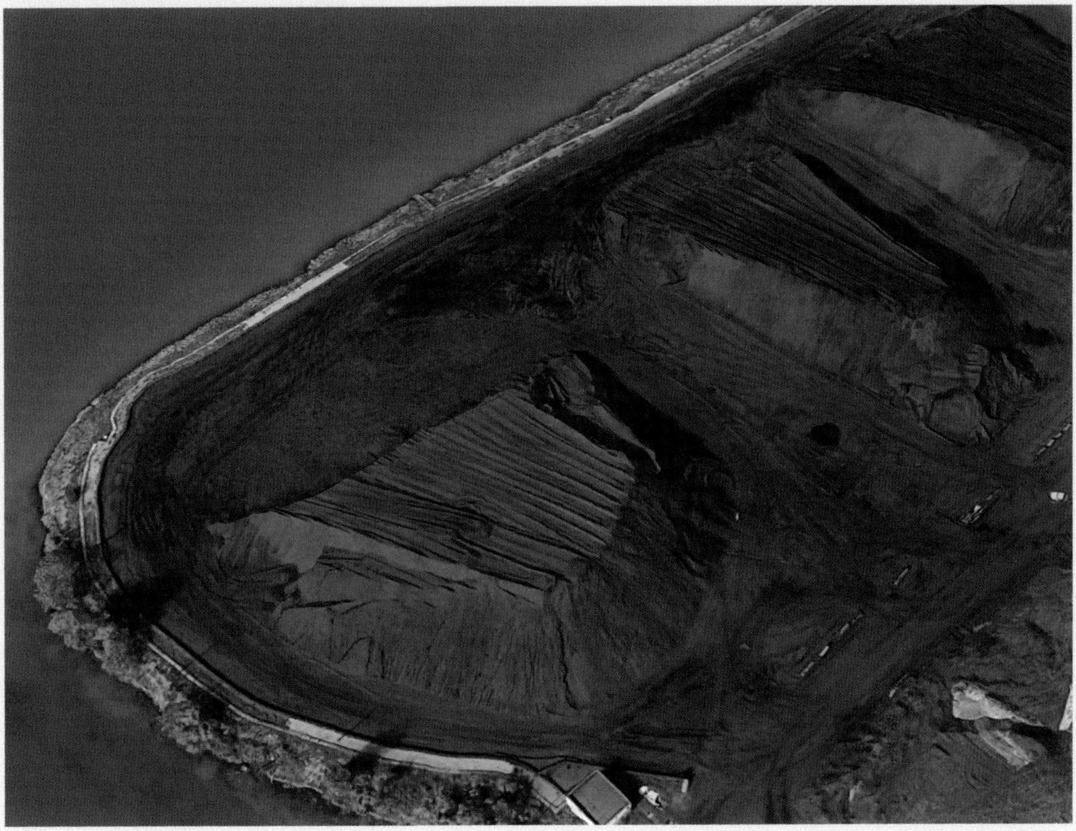

LOCALITY: Zug Island, Rouge River, Michigan, United States

COORDINATES 42.2842052, -83.1127004

LANSFORDITE

DESCRIPTION: Lansfordite was first discovered in 1888, within the slag dump of a coal mine in Lansford, Pennsylvania, United States. The mineral typically occurs as colorless to white, and can morph into crystals or stalactitic forms.

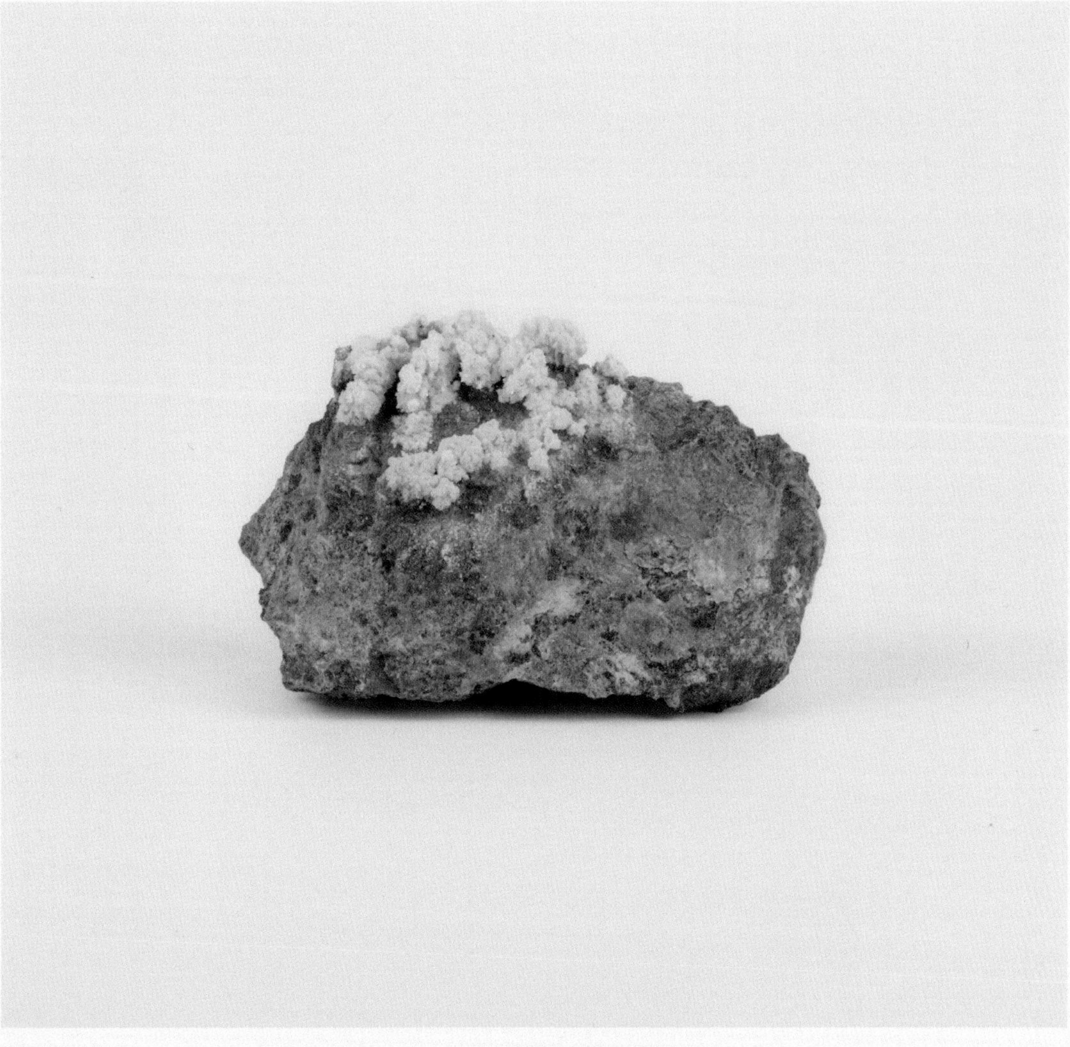

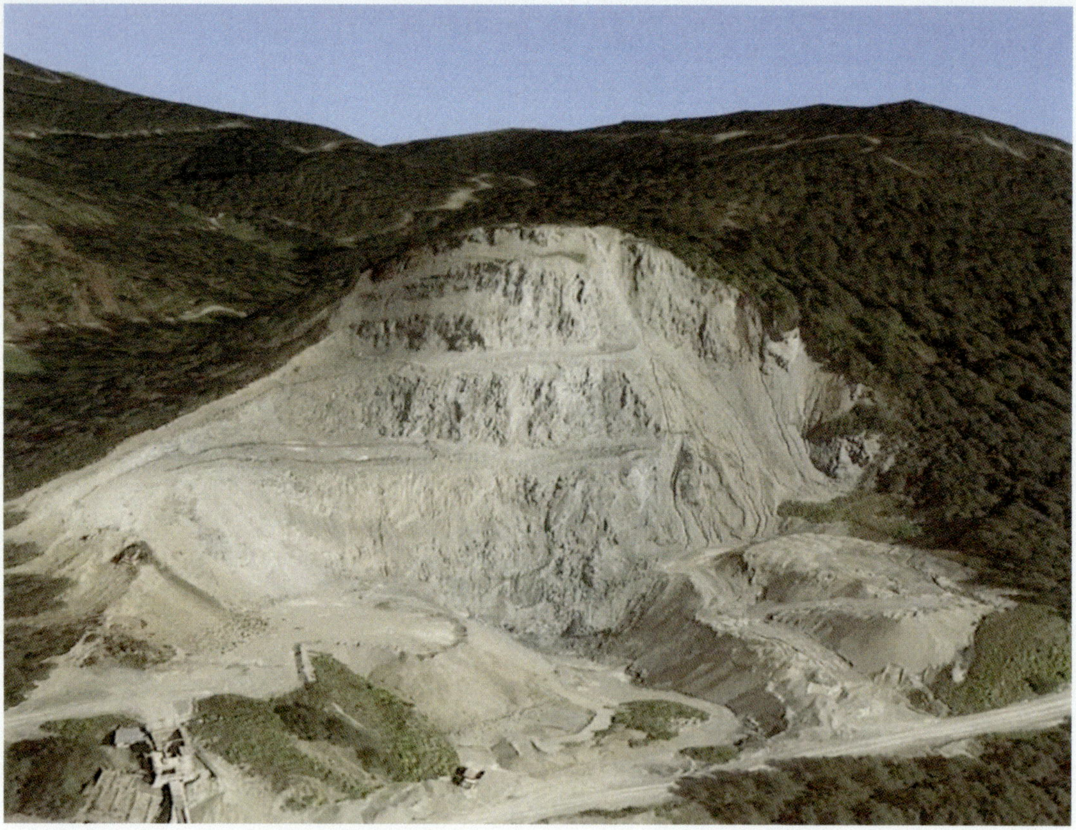

LOCALITY:	Sommingraben, Kraubath, Leoben, Styria, Austria
COORDINATES	47.29842, 15.01226

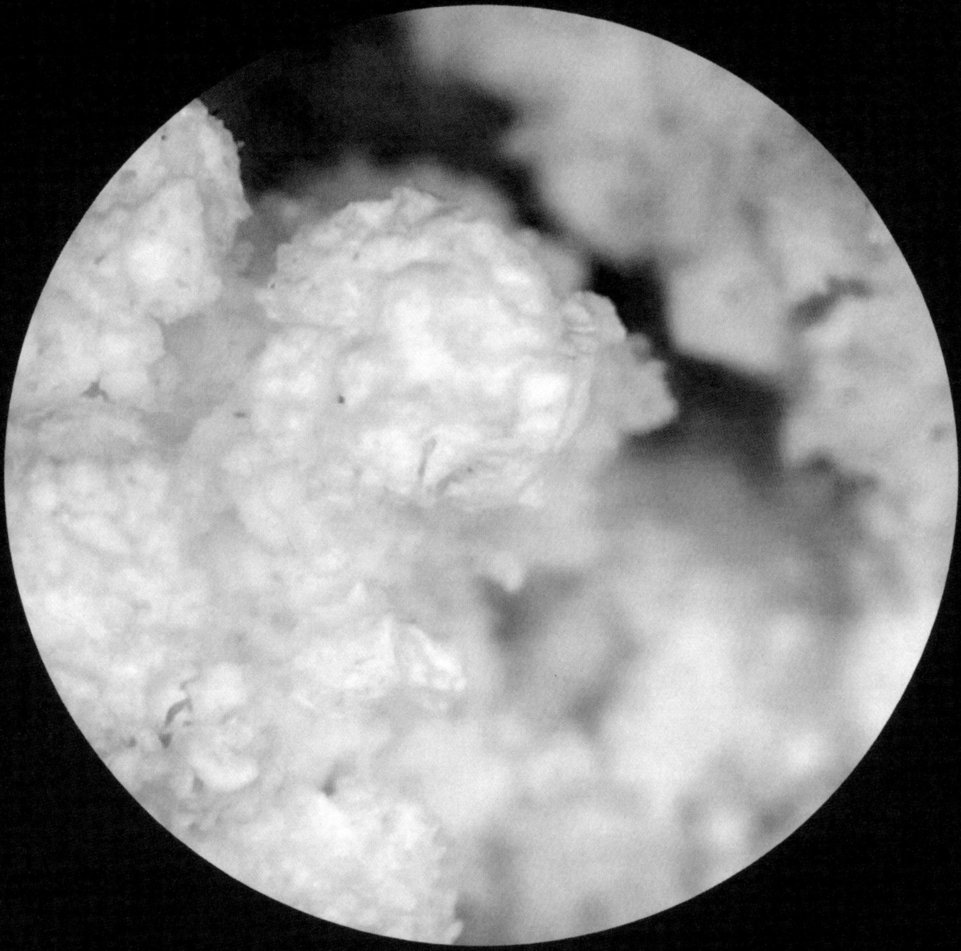

$MgCO_3 \cdot 5H_2O$

METAMUNIRITE

DESCRIPTION:

Metamunirite forms when munirite undergoes wetting and drying processes. Extremely rare, the mineral is found in the cavities in vanadium-bearing sandstone. This sample was found in the Sunday Mine Complex located in San Miguel County, Colorado. The Sunday Mine Complex currently consists of the only fully operational uranium and vanadium mines in the United States.

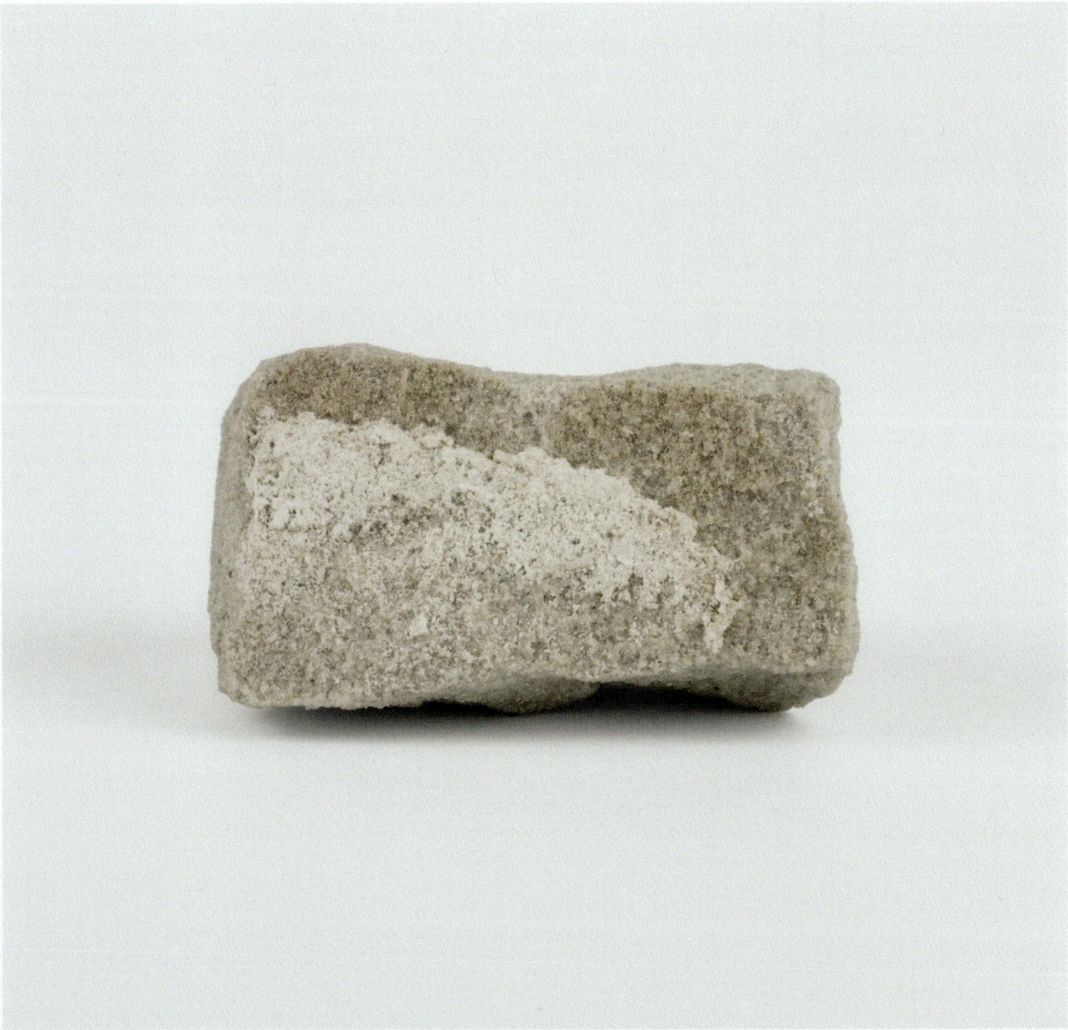

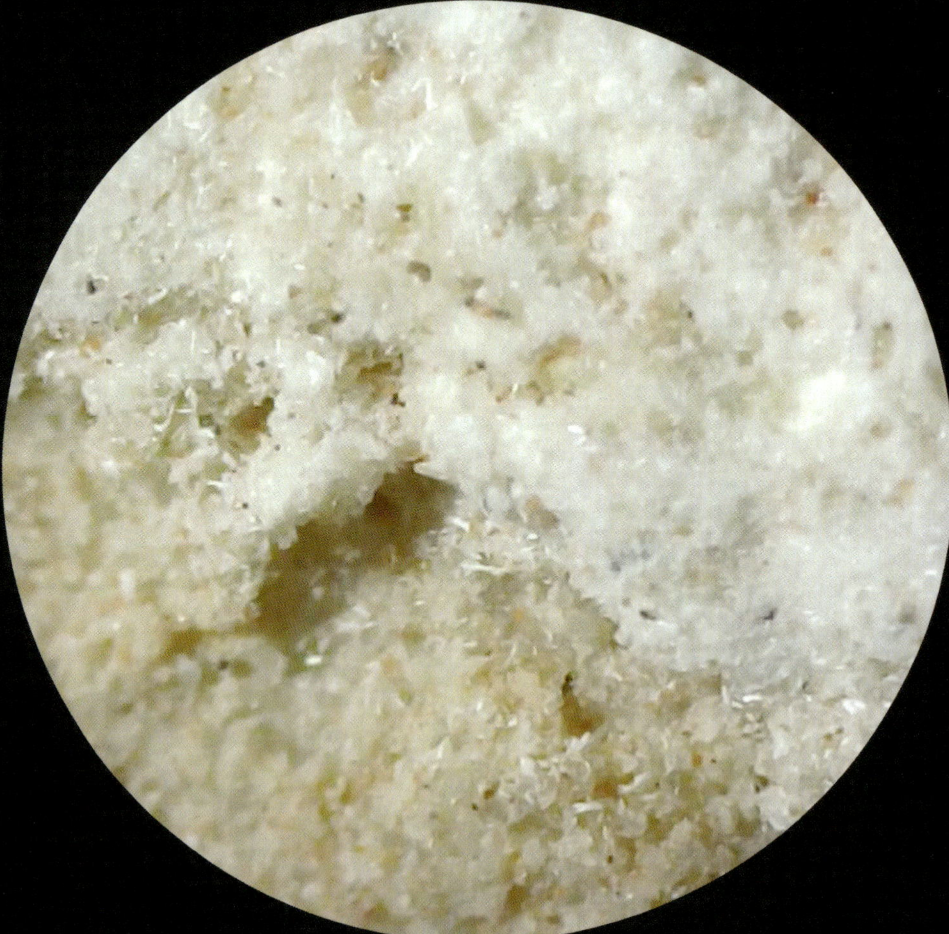

$NaVO_3$

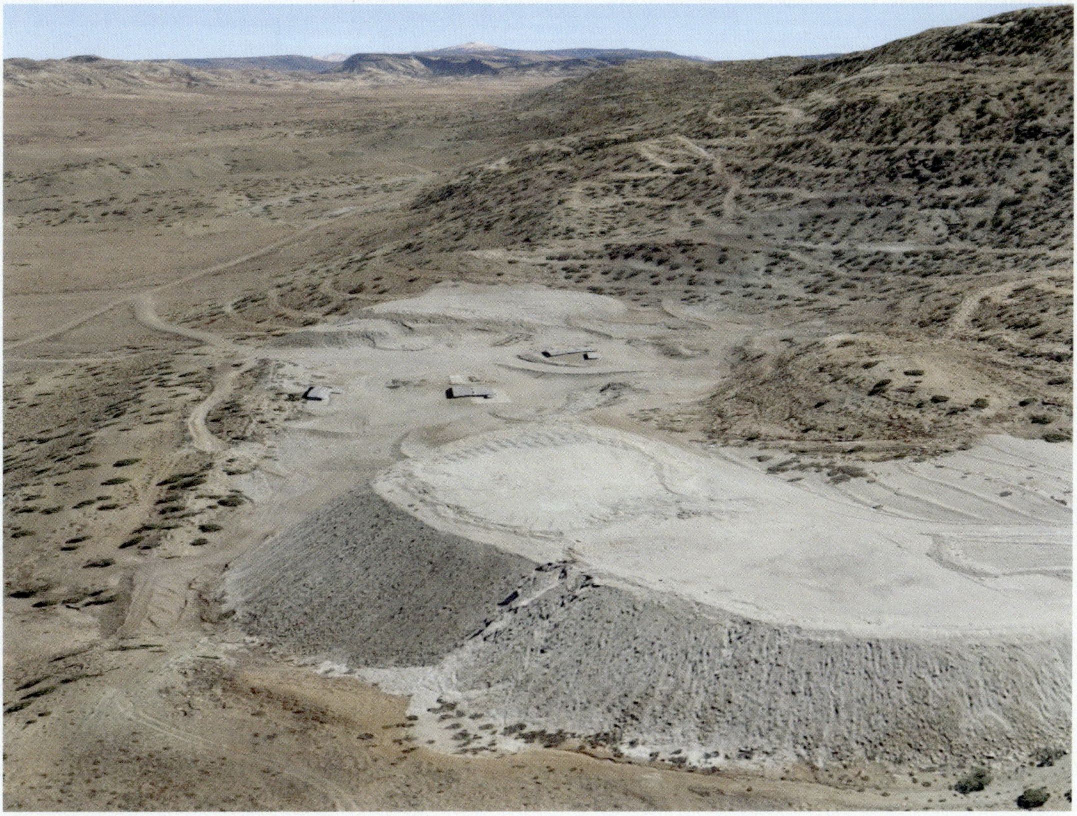

LOCALITY:	Sunday Mine, Slick Rock Mining District, San Miguel County, Colorado, United States
COORDINATES	38.07190,-108.80487

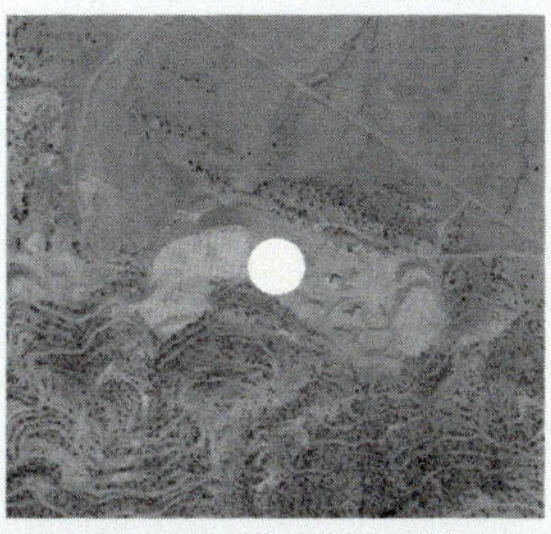

NESQUEHONITE

DESCRIPTION:

Nesquehonite, named after its city of origin, was discovered in a coal mine in Nesquehoning, Pennsylvania, United States. The mineral forms during the wetting and drying cycles that occur within ultramafic mining waste piles. Nesquehonite has been the subject of recent ecological experiments, including attempts to use it as a new "green" building material, and as a resource to offset the energy usage of cryptocurrency mining.

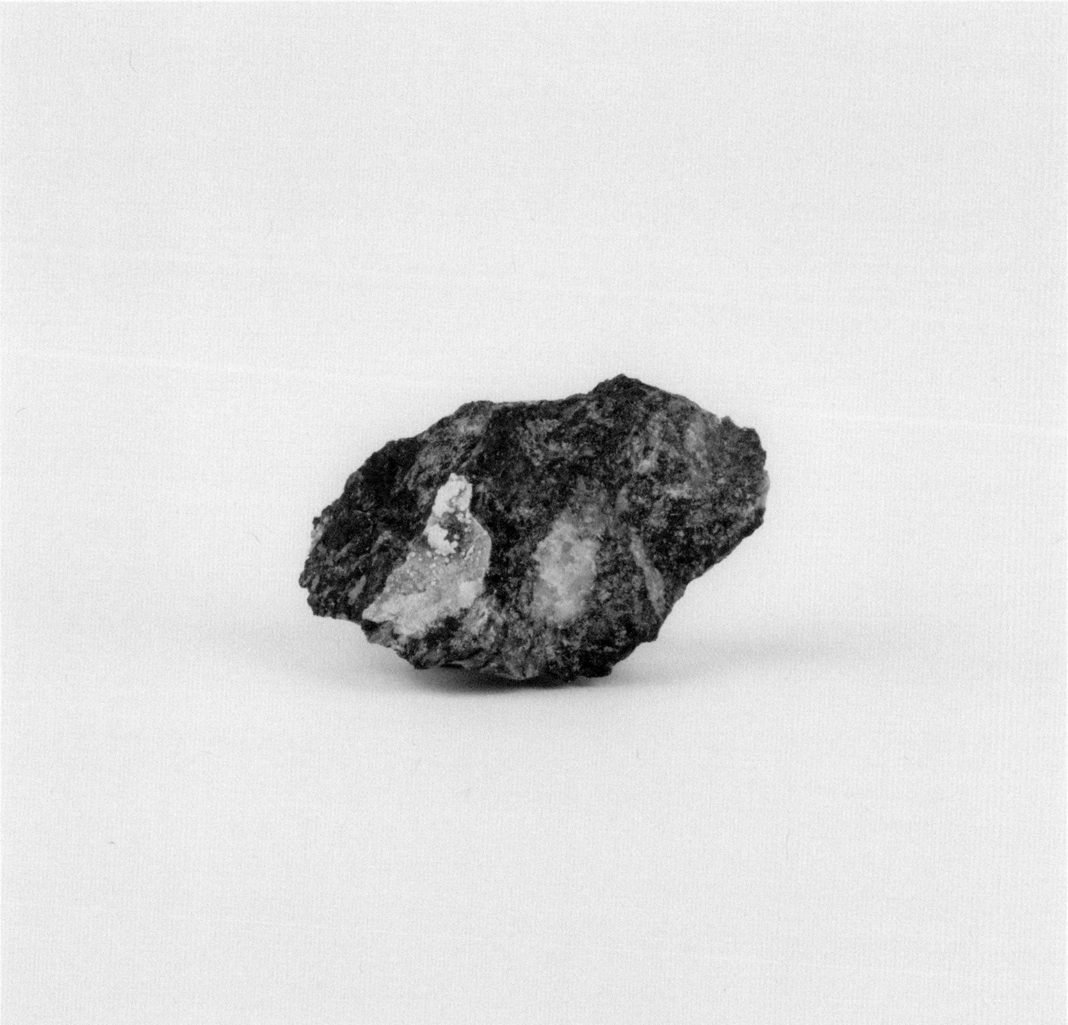

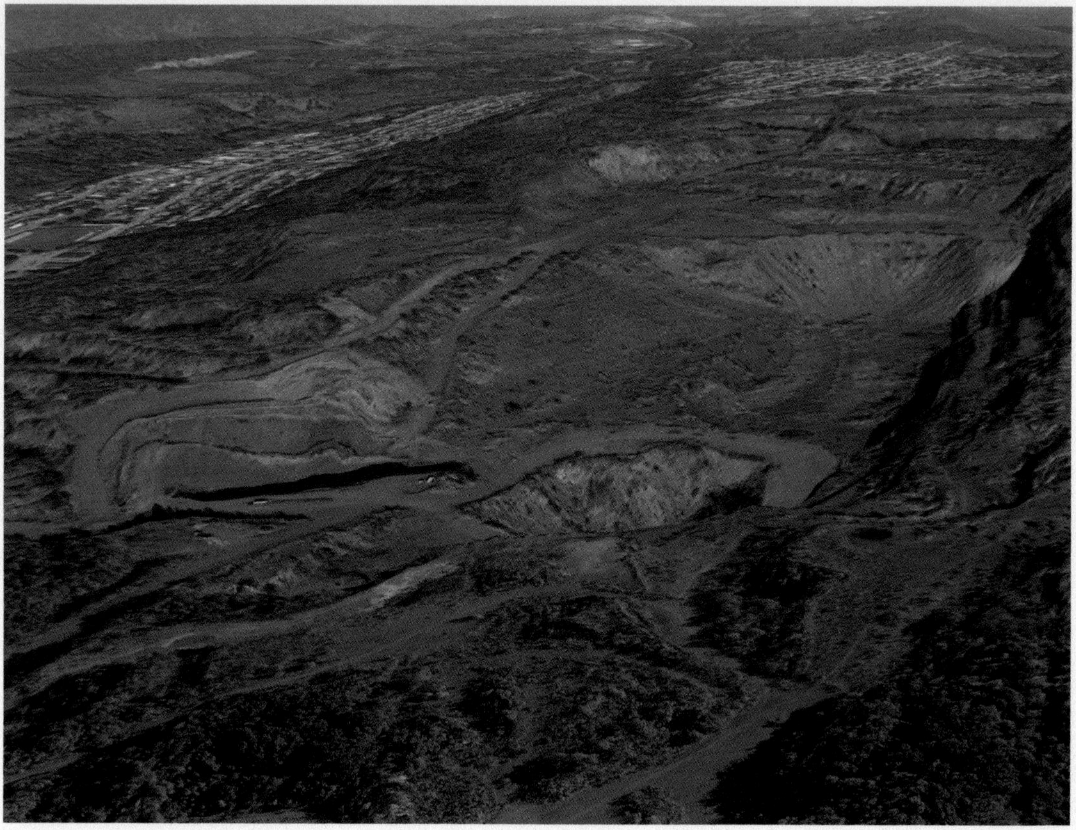

LOCALITY: Nesquehoning, Carbon County, Pennsylvania, United States

COORDINATES 40.86250, -75.81250

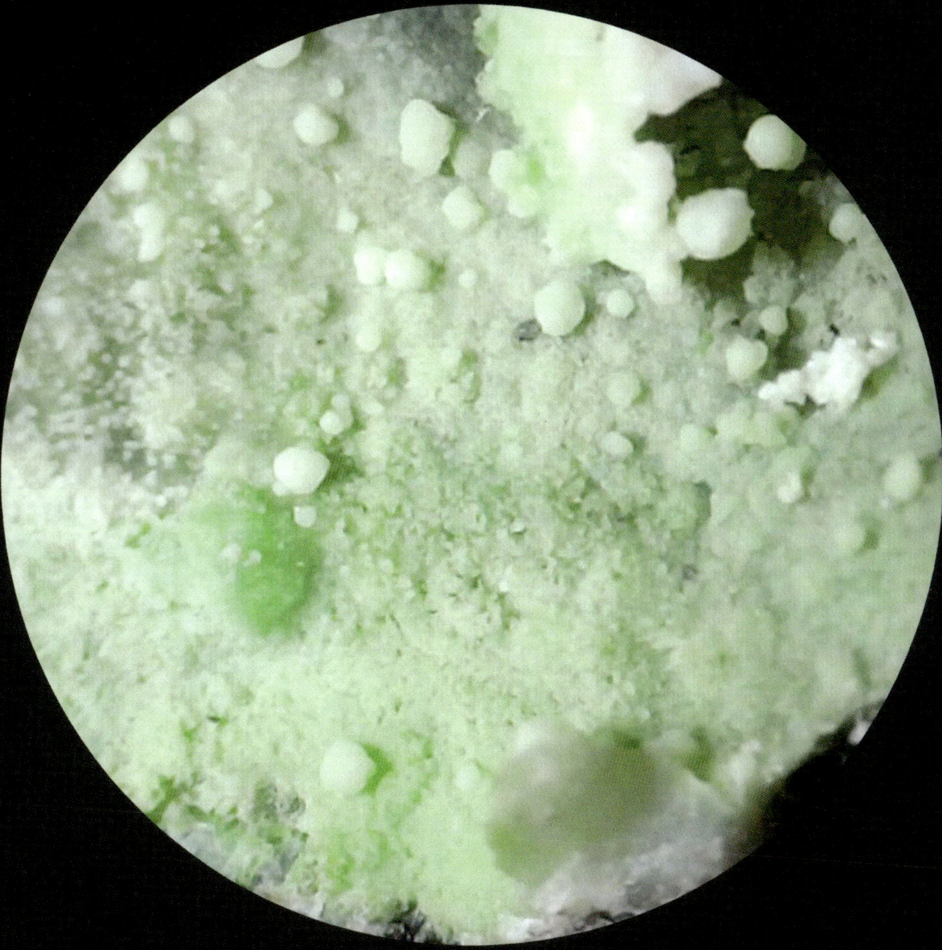

$MgCO_3 \cdot 3H_2O$

PLASTIGLOMERATE

DESCRIPTION:

Plastiglomerates form along ocean shorelines where natural sedimentary grains and other organic debris (shells, wood, etc.) are fused together by melting plastic during campfires. They have been primarily discovered on the Hawaiian islands, where volcanic sediment is present. It is closely related to Plastistone, a stone made almost completely of plastic, which was discovered on the volcanic islands of Brazil.

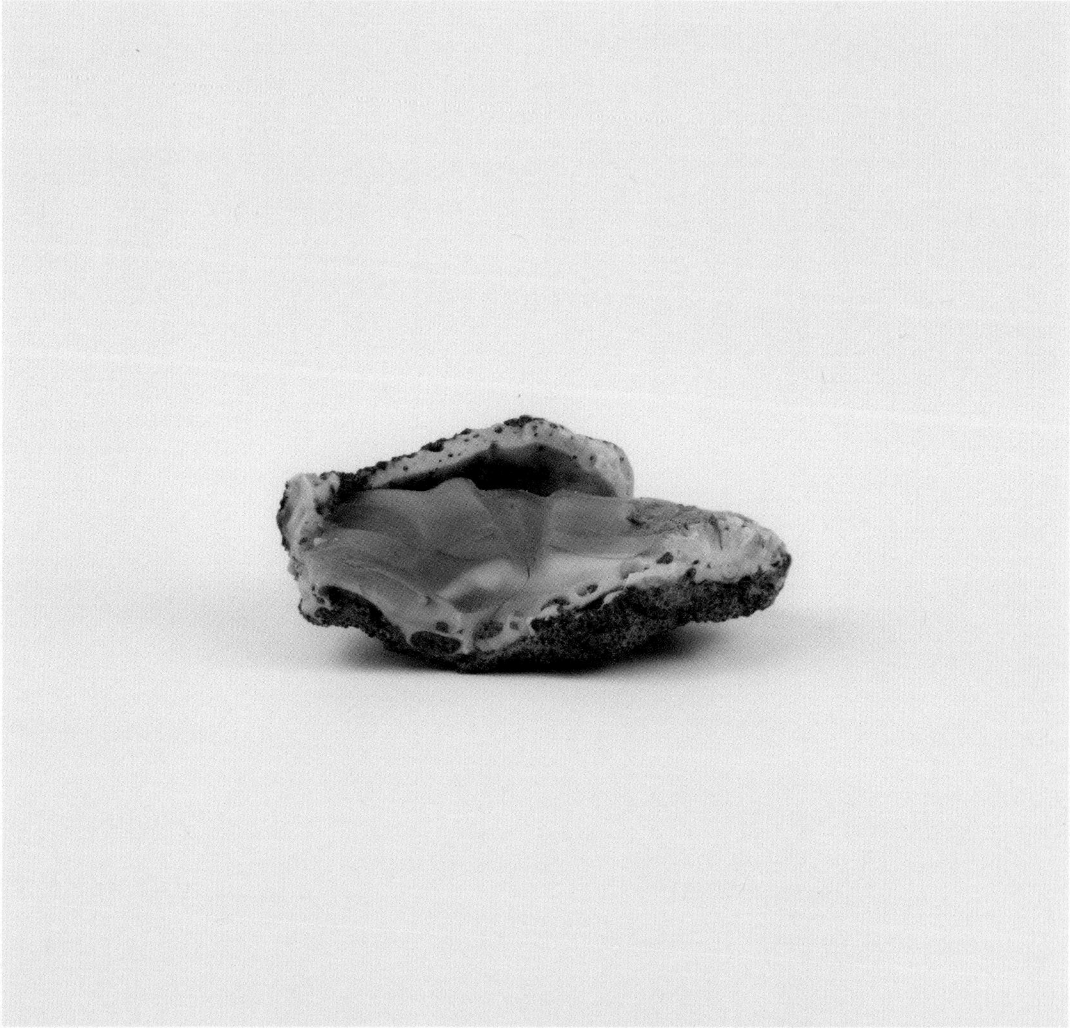

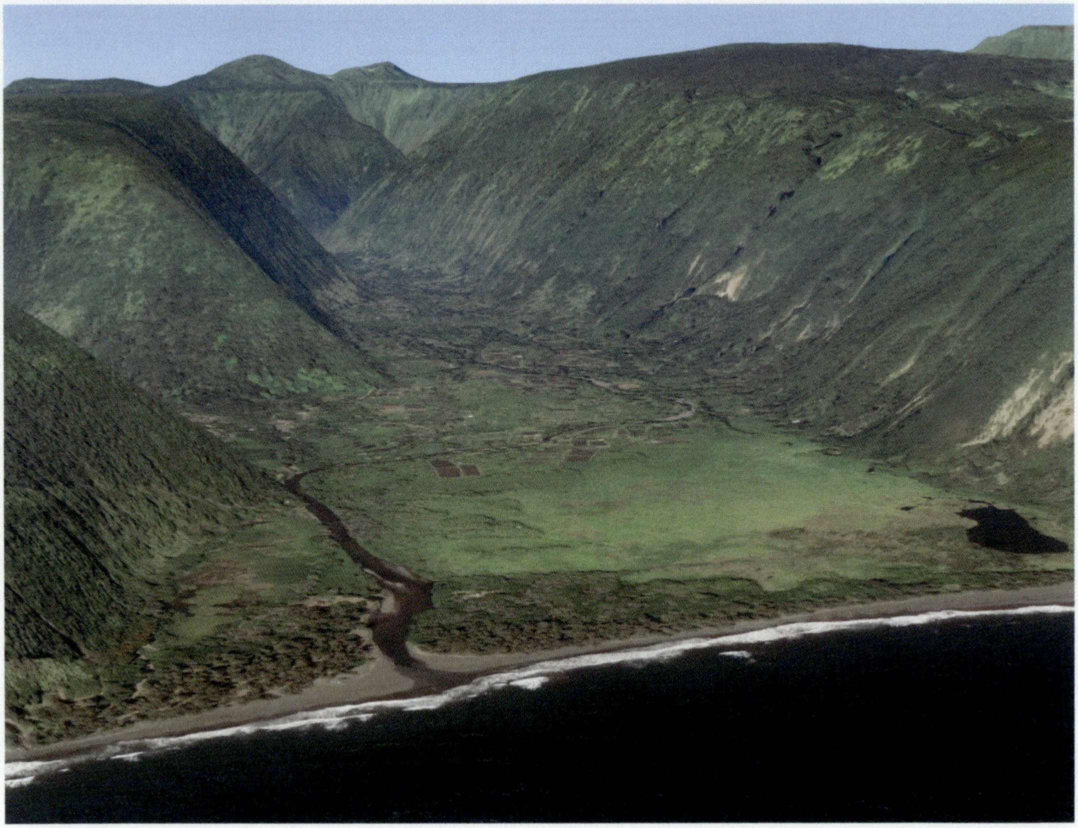

LOCALITY: Waipi'o Valley Black Sand Beach, Honokaa, Hawaii, United States

COORDINATES 20.11862, -155.58874

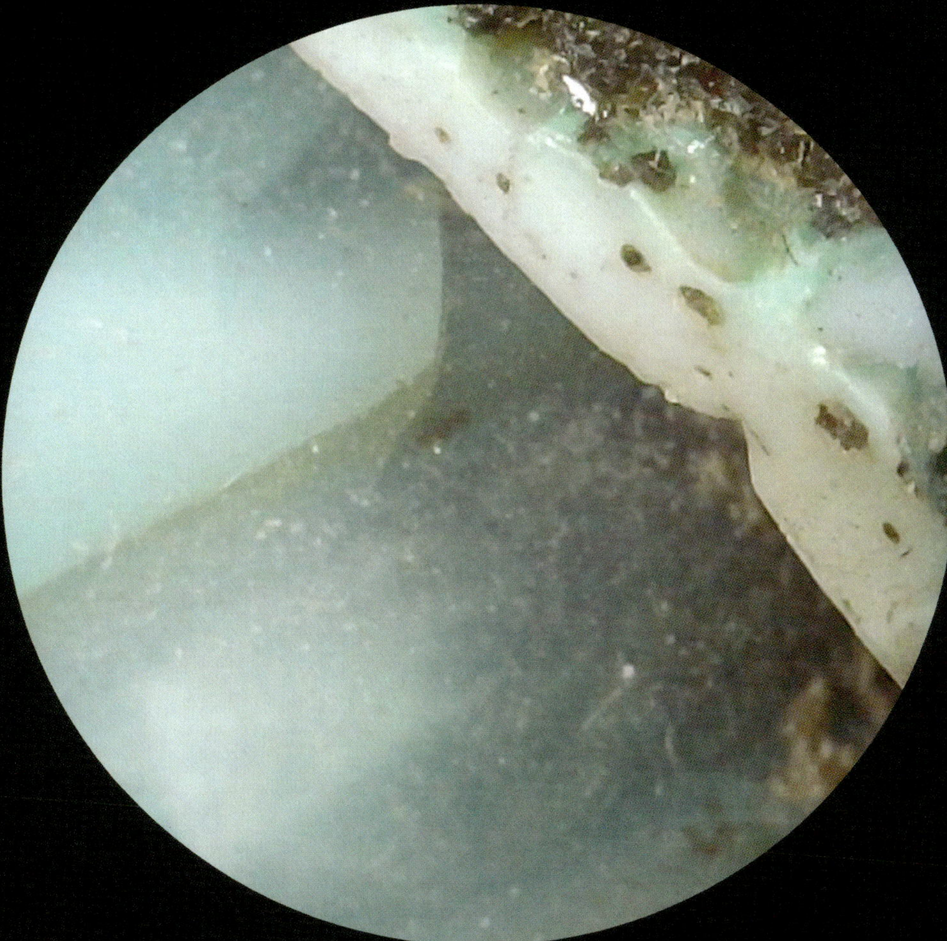

CHEMICAL COMPOSITION VARIES

PHOSGENITE

DESCRIPTION:

Phosgenite was first discovered in 1820 in the altered cavities of a lead mine in Cromford, England, which is known as the "birthplace of the industrial revolution." The mineral is created through a post-mining oxidation of lead and slag when it comes into prolonged contact with sea water. This particular sample comes from the ancient mines of Lavrion in Greece, which was known for extracting silver, copper, and lead.

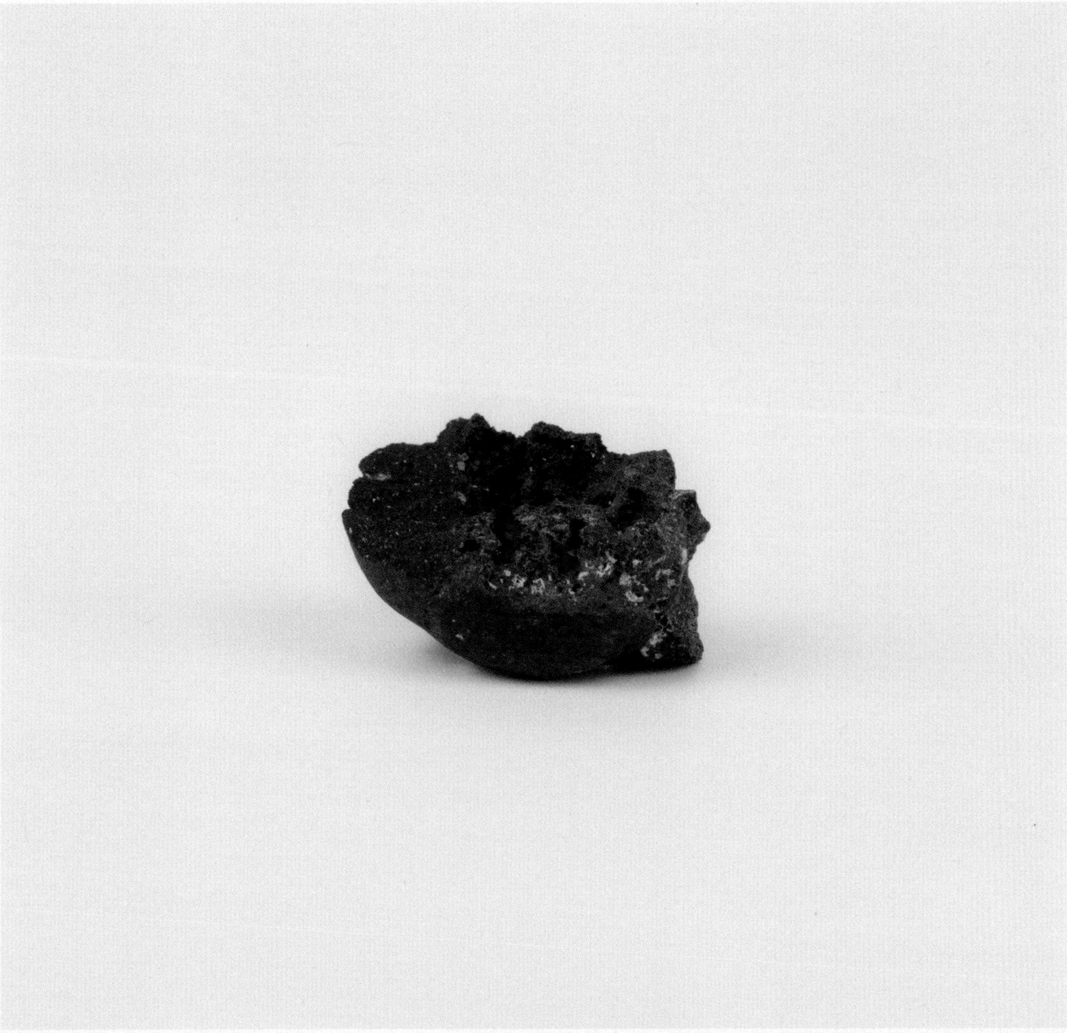

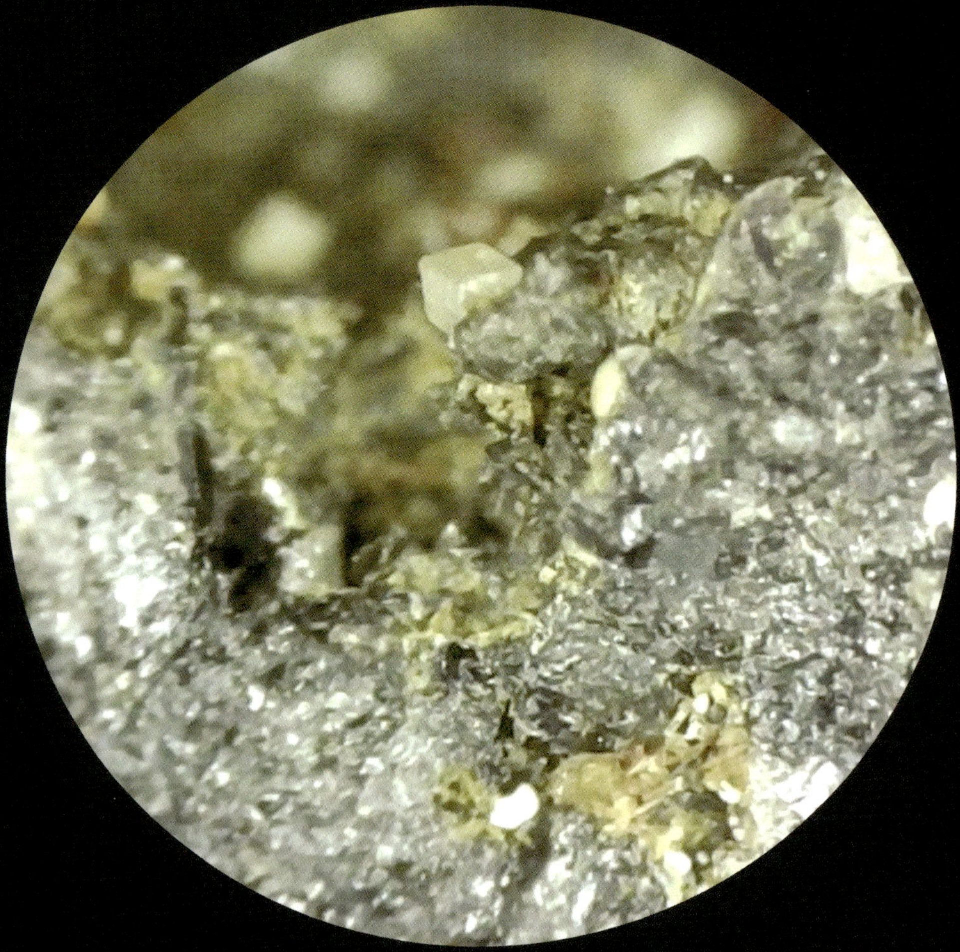

$Pb_2CO_3Cl_2$

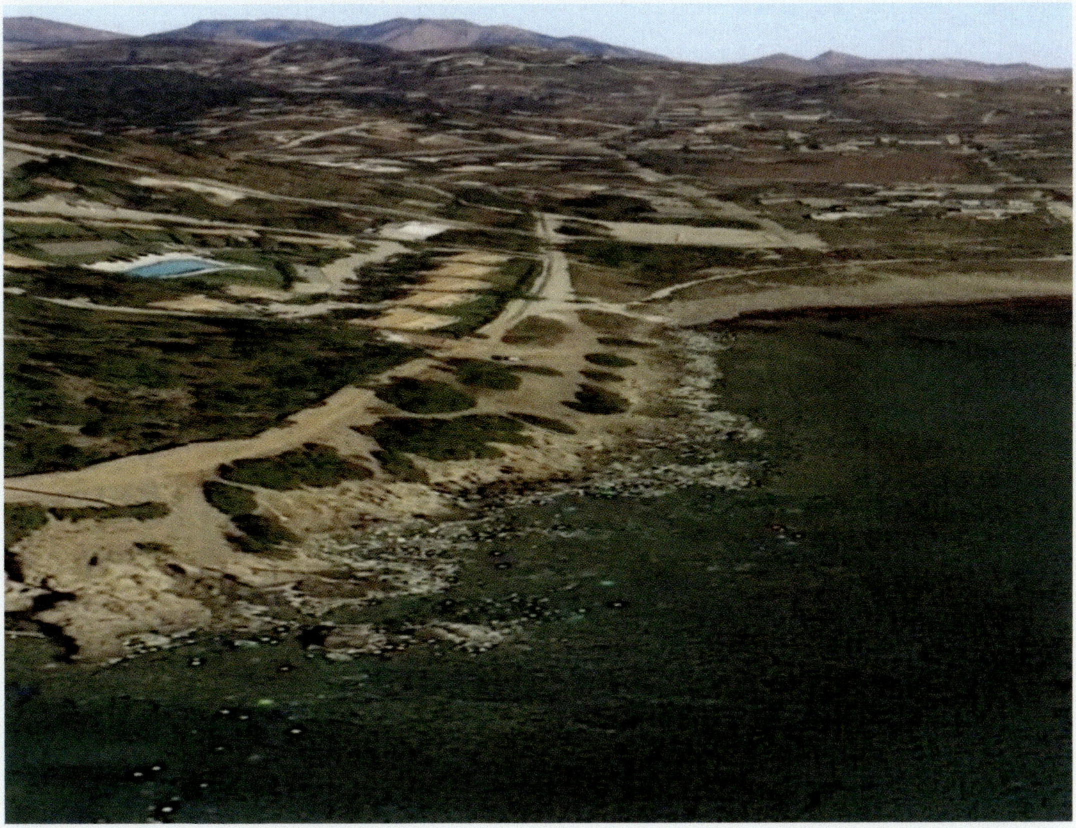

LOCALITY: Passa Limani Cove, Lavrion District, Attiki Prefecture, Greece

COORDINATES 37.67681, 24.05180

RAINBOW SLAG

DESCRIPTION:

Rainbow Slag, sometimes called Peacock Coal, is named for the rainbow effect on the surface of the mineral. It is caused when sulfur oxidizes on the surface of coal, creating an iridescent film. This sample of Rainbow slag originates from the coal mines on Zug Island in Michigan, United States.

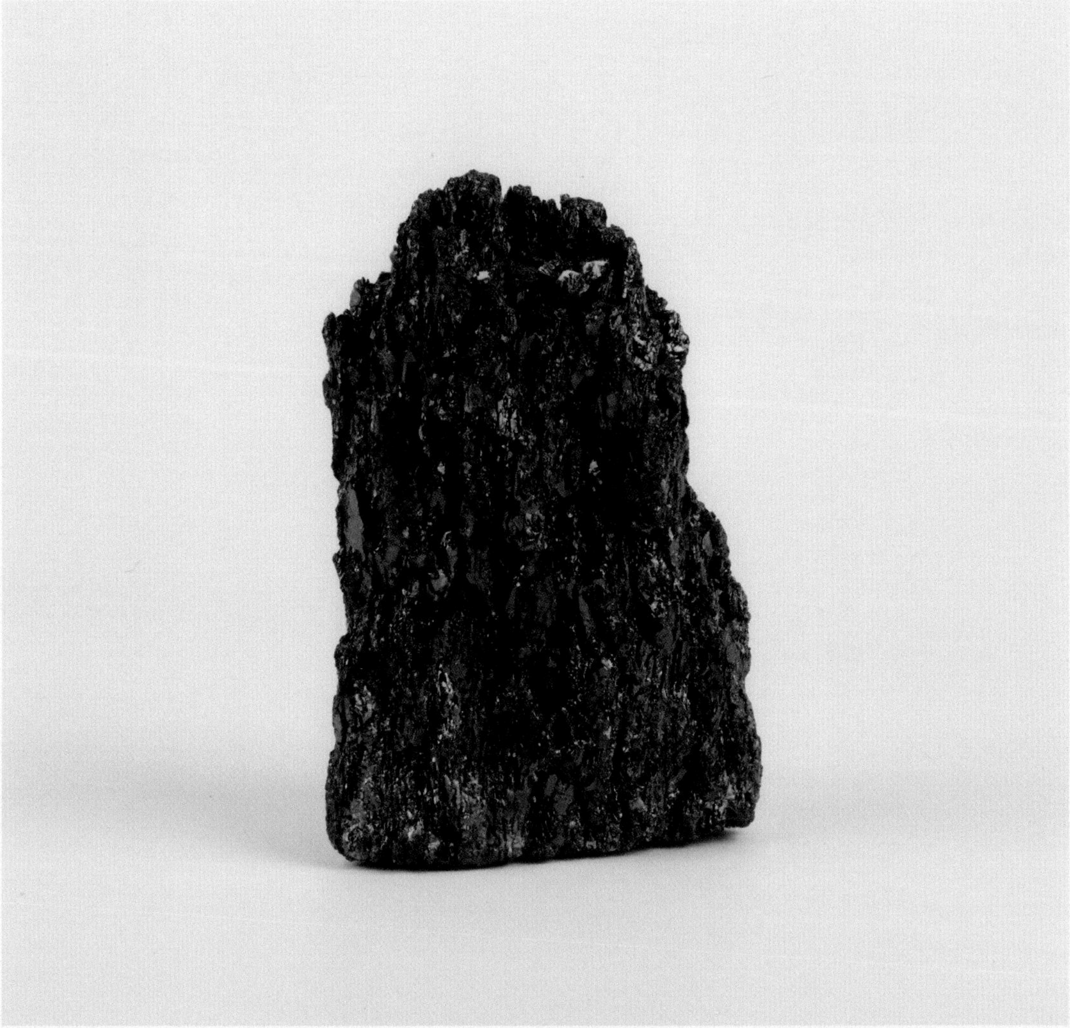

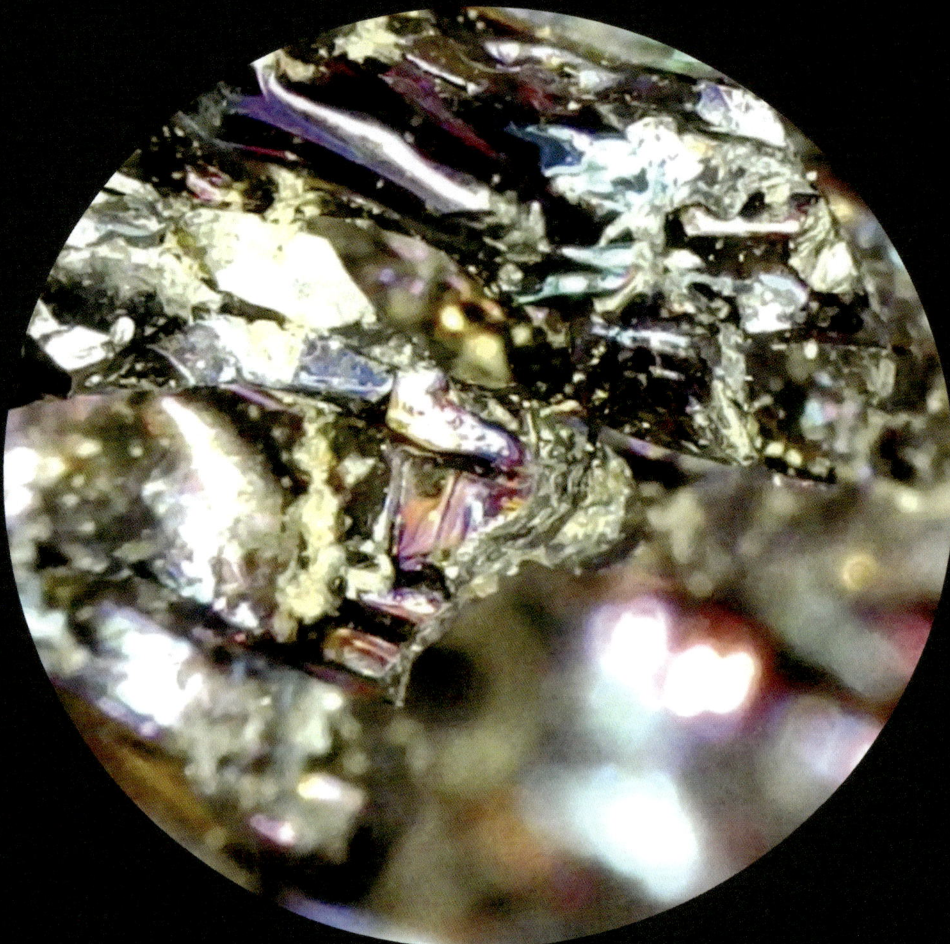

$C_{240}H_{90}O_4NS \cdot SO_2$

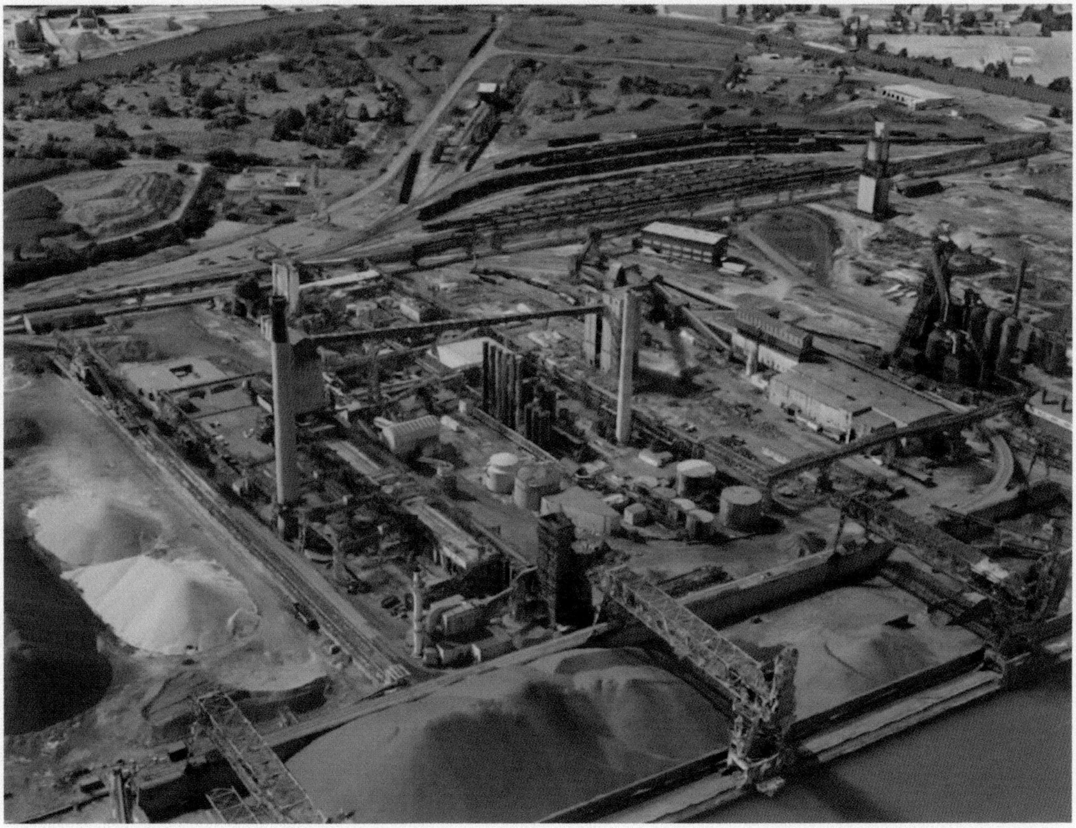

LOCALITY: Zug Island, Rouge River, Michigan, United States

COORDINATES 42.2842052, -83.1127004

SHANNONITE

DESCRIPTION:

Shannonite is a lead oxycarbonate that forms when cerussite is heated during fire-setting methods used in ore extraction, also known as mine fires. This piece was found in the lead and silver extraction site of the Tonopah-Belmont Mine in Arizona, United States.

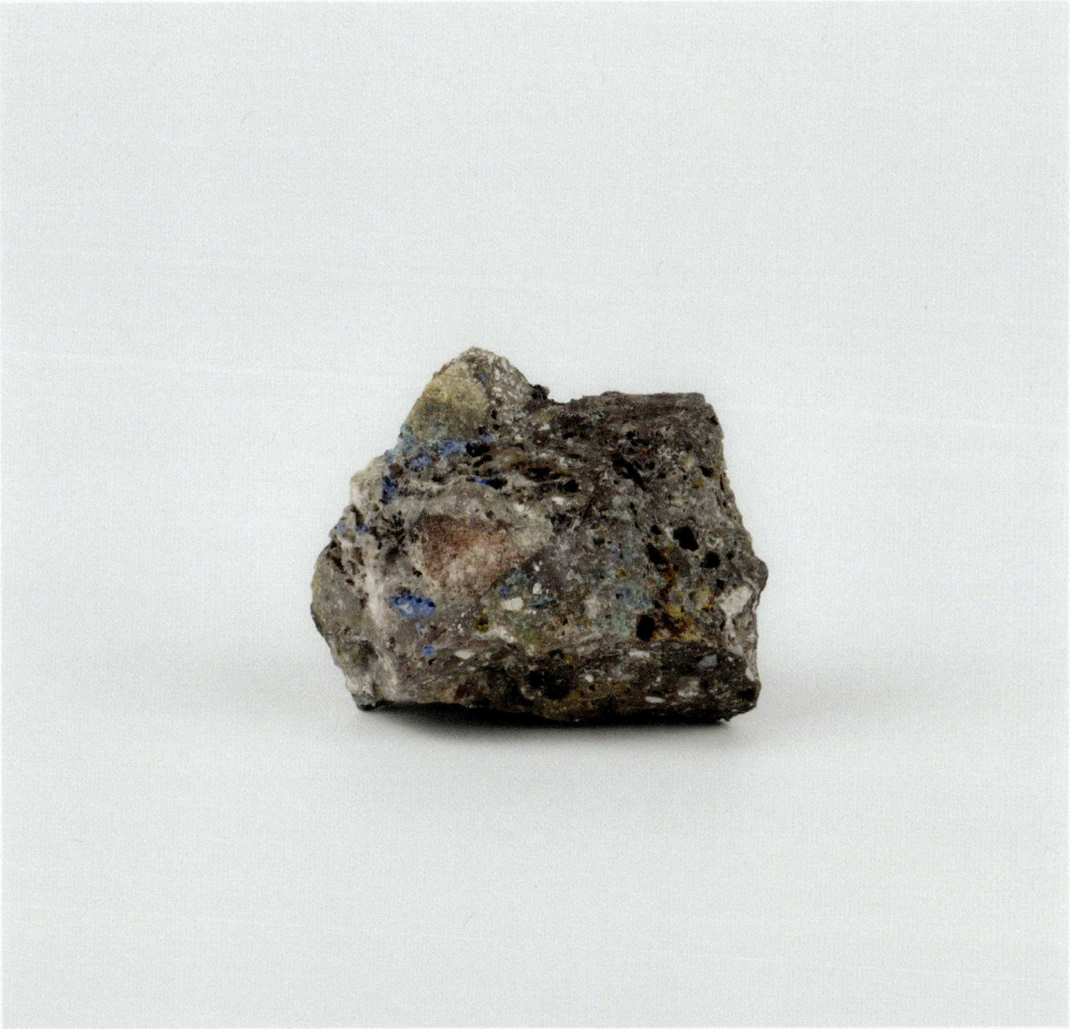

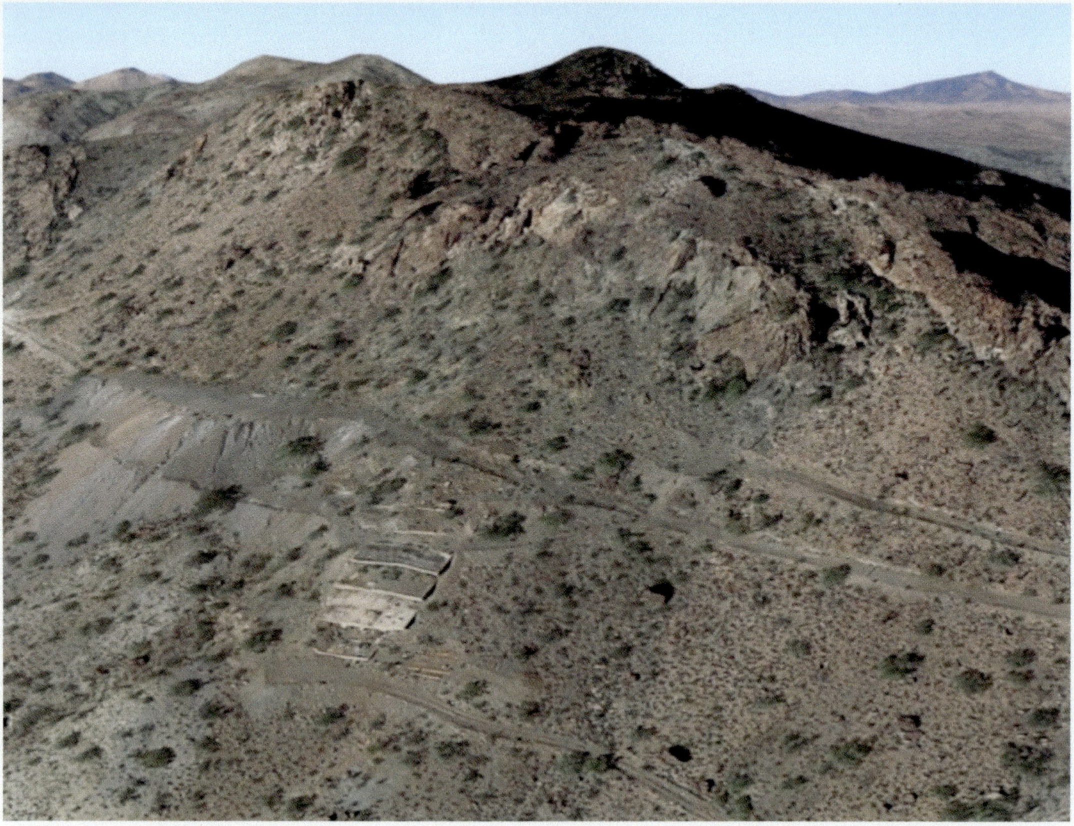

LOCALITY: Tonopah-Belmont Mine, Osborn, Maricopa County, Arizona, United States

COORDINATES 33.64778, -112.93222

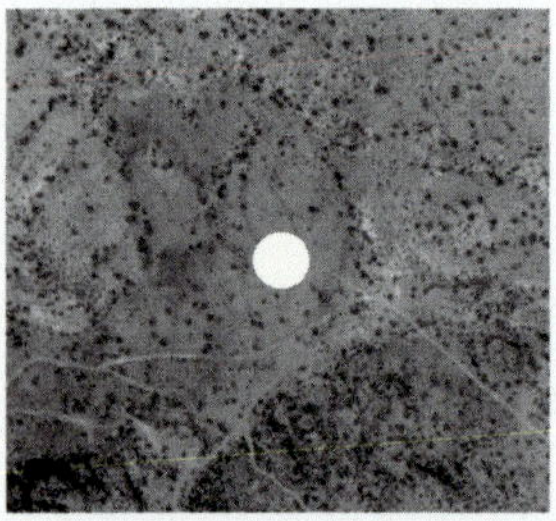

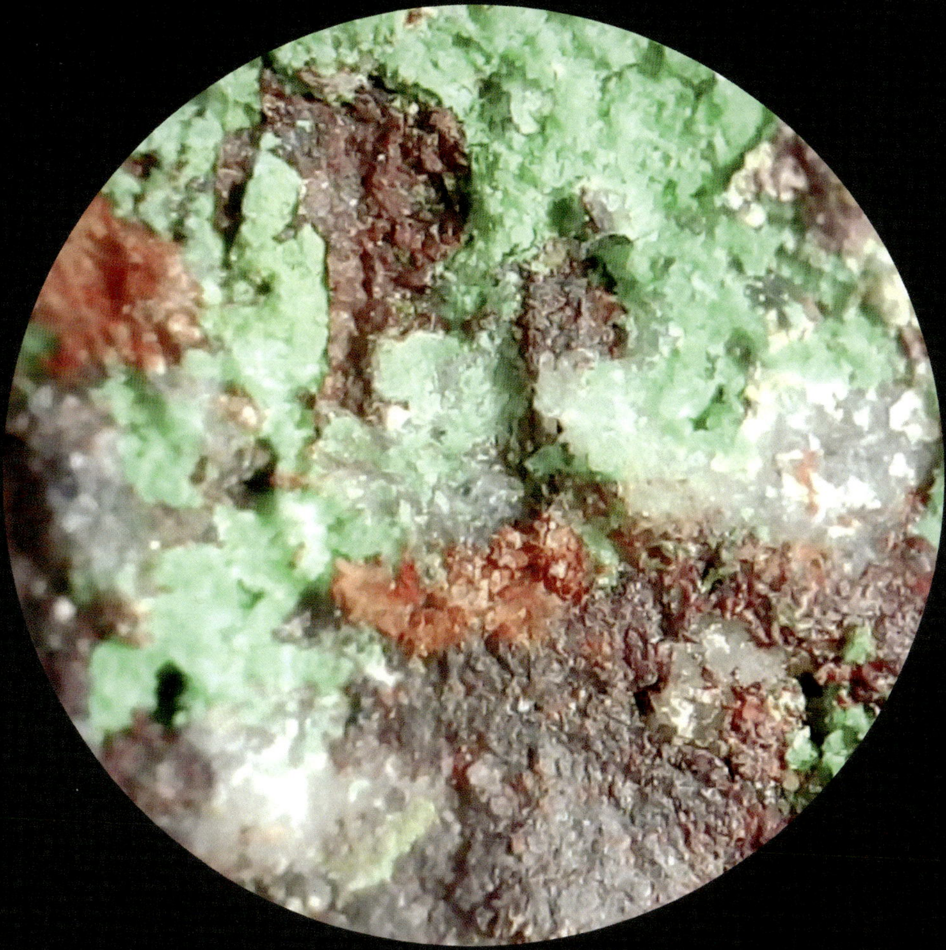

Pb_2OCO_3

SIMONKOLLEITE

DESCRIPTION:

Simonkolleite is a rare mineral formed by the weathering and corrosion of zinc-bearing slag. The mineral has shown to be a desirable nutritional supplement for livestock and is also used within fungicide. This sample was found in the slag dumps of an ancient copper smelter in Hesse, Germany.

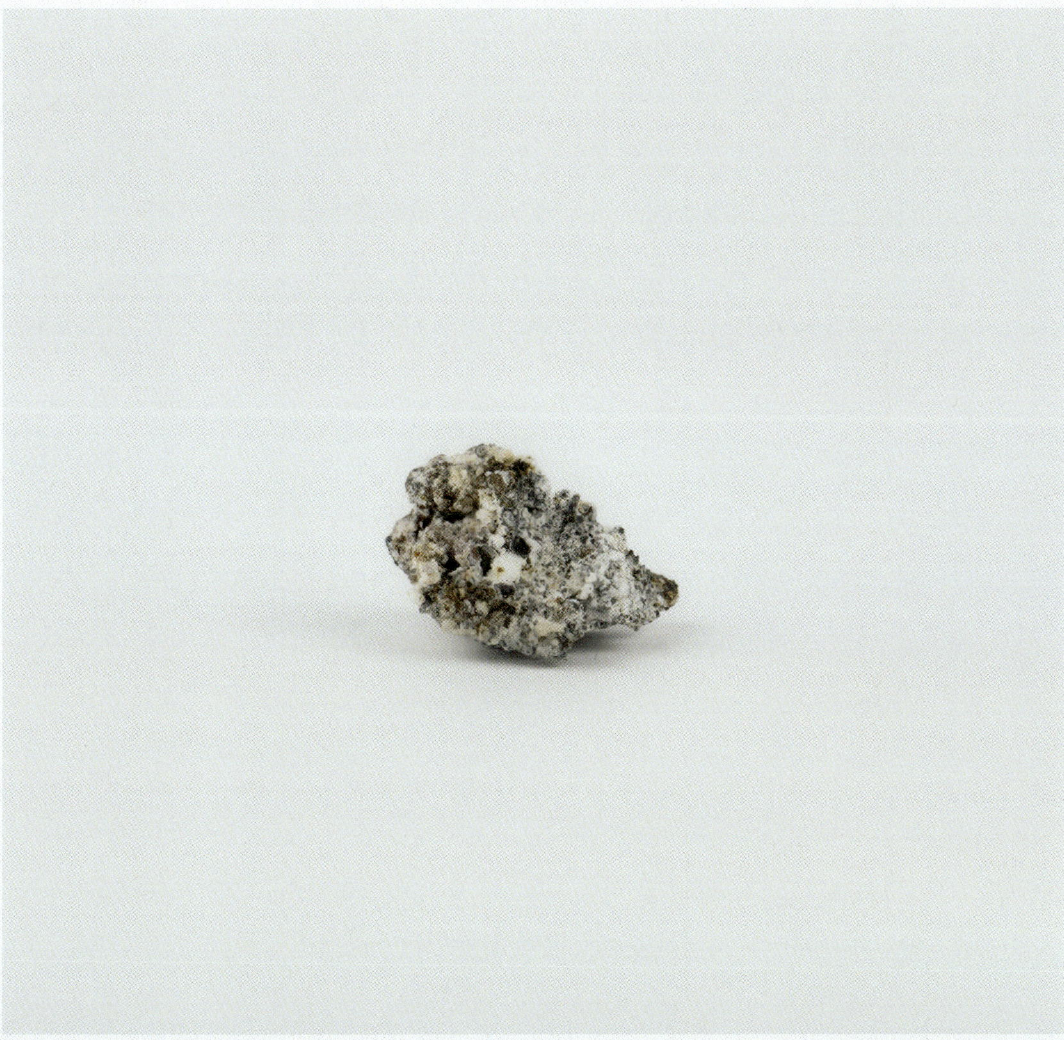

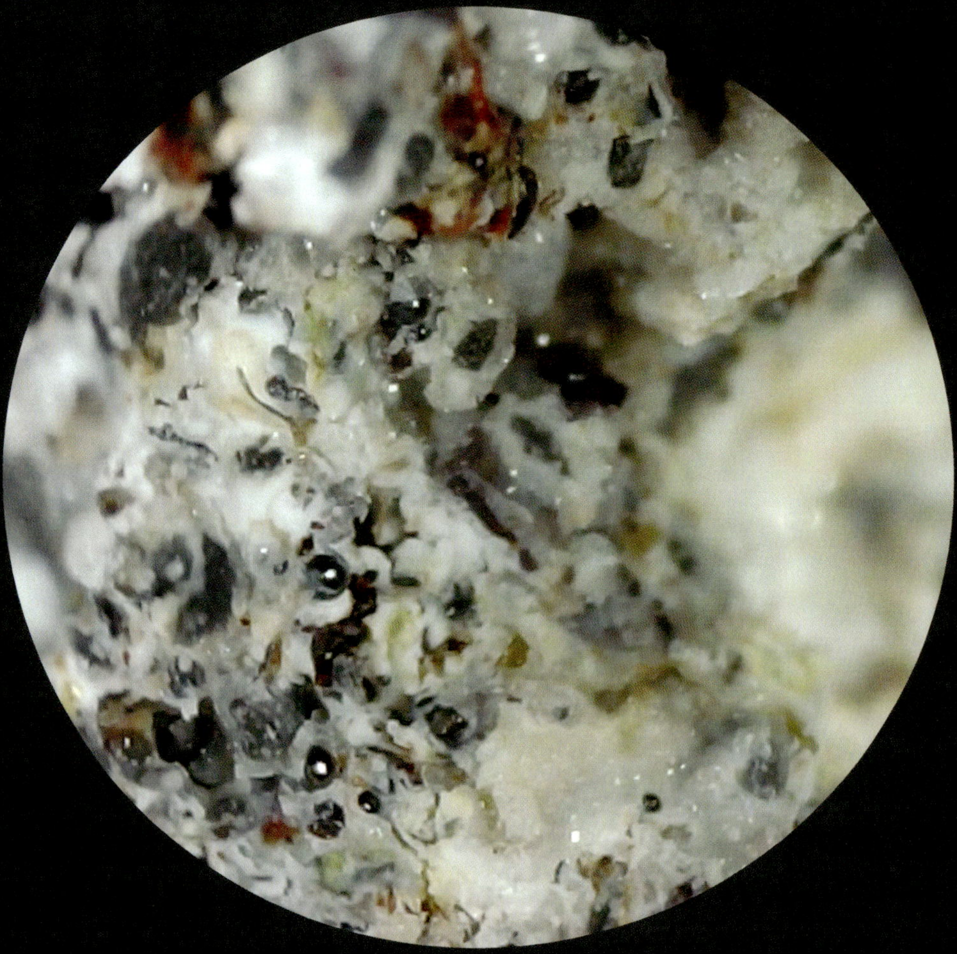

$Zn_5Cl_2(OH)_8 \cdot H_2O$

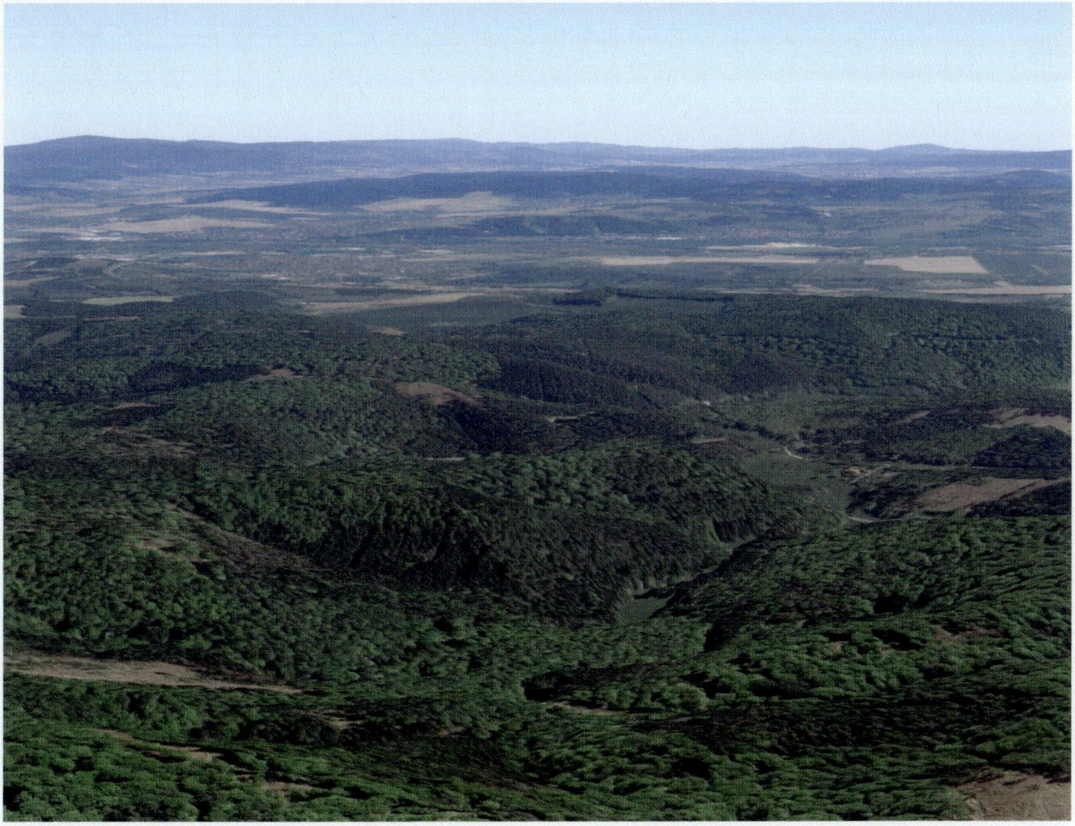

LOCALITY:	Richelsdorf Smelter, Süß, Nentershausen, Hersfeld-Rotenburg, Kassel Region, Hesse, Germany
COORDINATES	50.98944, 9.97310

TINNUNCULITE

DESCRIPTION: Tinnunculite forms when hot gases escape from burning coal dumps and react with excrement from Falco tinnunculus, or the common kestrel bird.

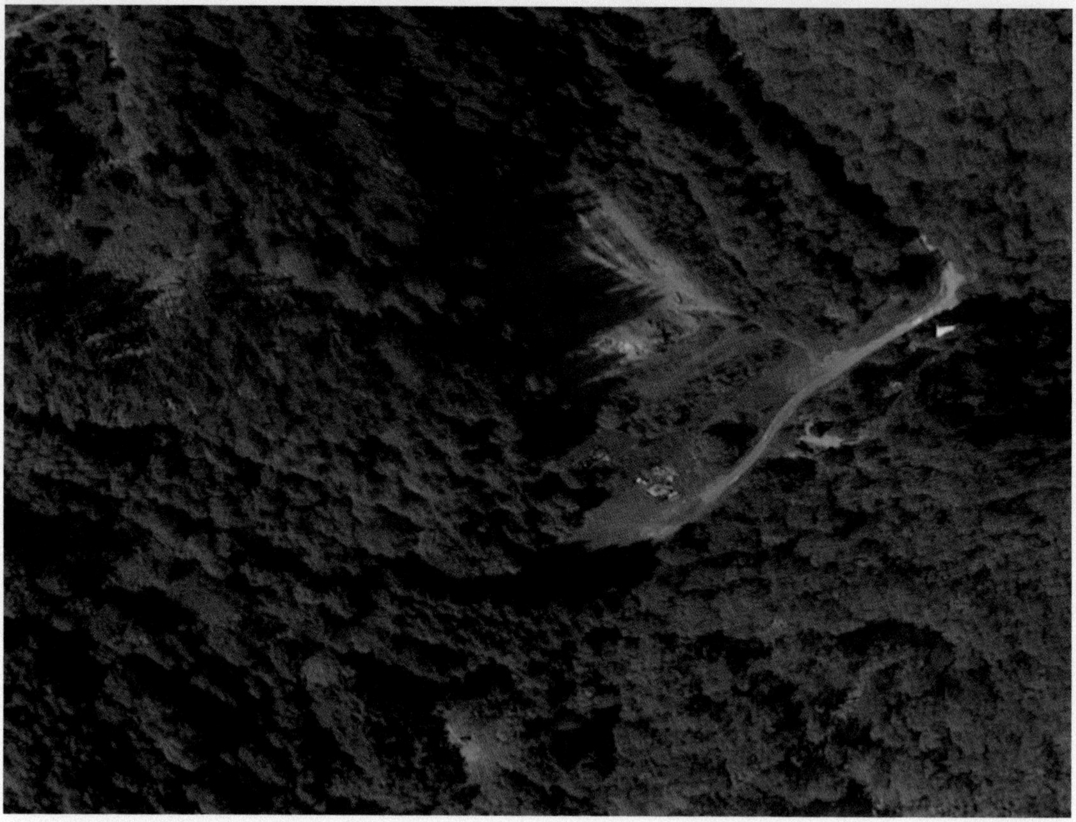

LOCALITY:	Adolf-Helene Mine, Altlay, Zell, Cochem-Zell, Rlp, Germany
COORDINATES	49.994720, 7.285560

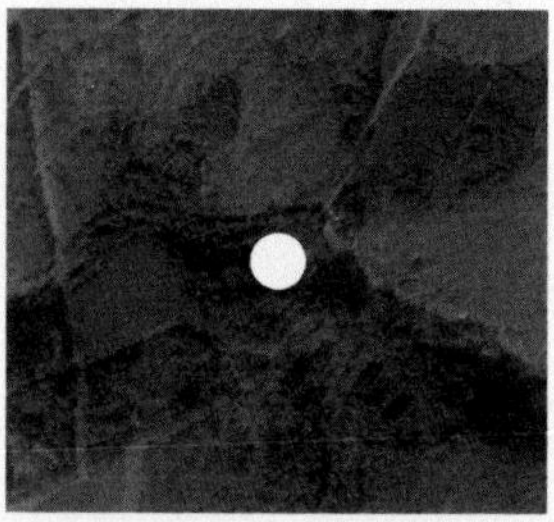

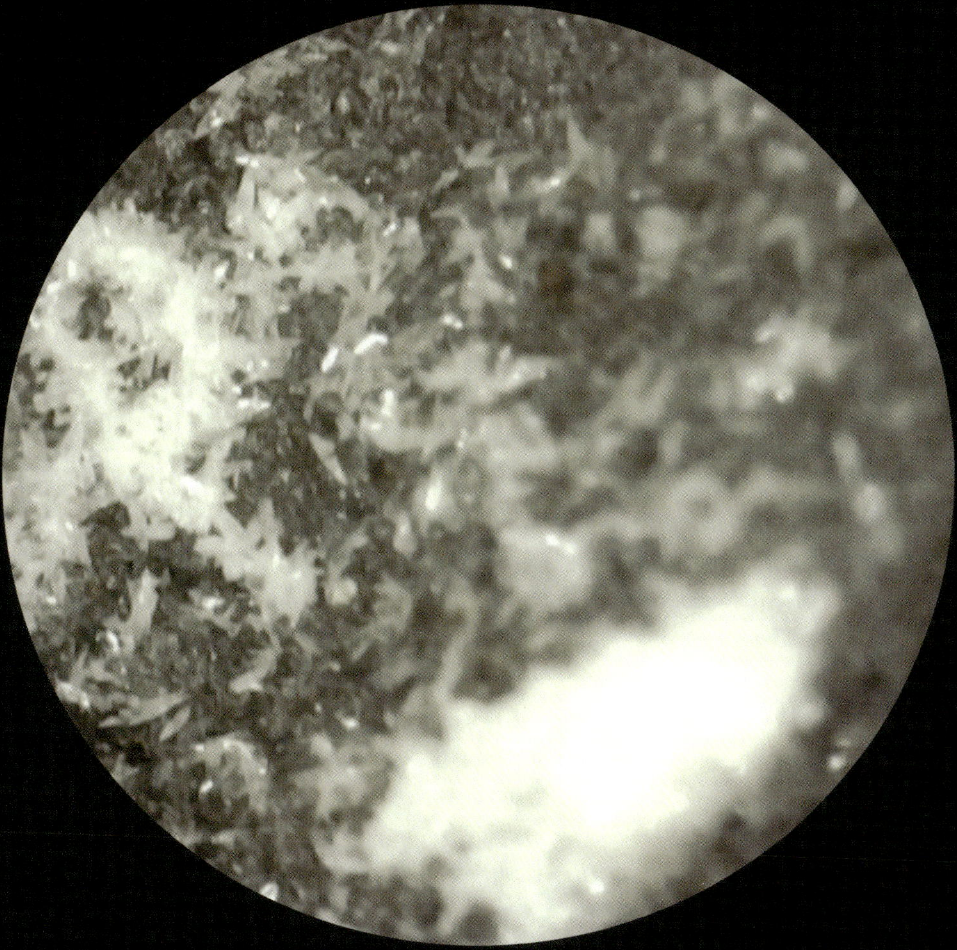

$C_5H_4N_4O_3 \cdot 2H_2O$

TRINITITE

DESCRIPTION: Trinitite is the glassy residue which was left on the desert floor after the plutonium-based Trinity nuclear bomb test took place in New Mexico, United States on July 16th, 1945. It is composed of arkosic sand and feldspar that was melted by the atomic blast.

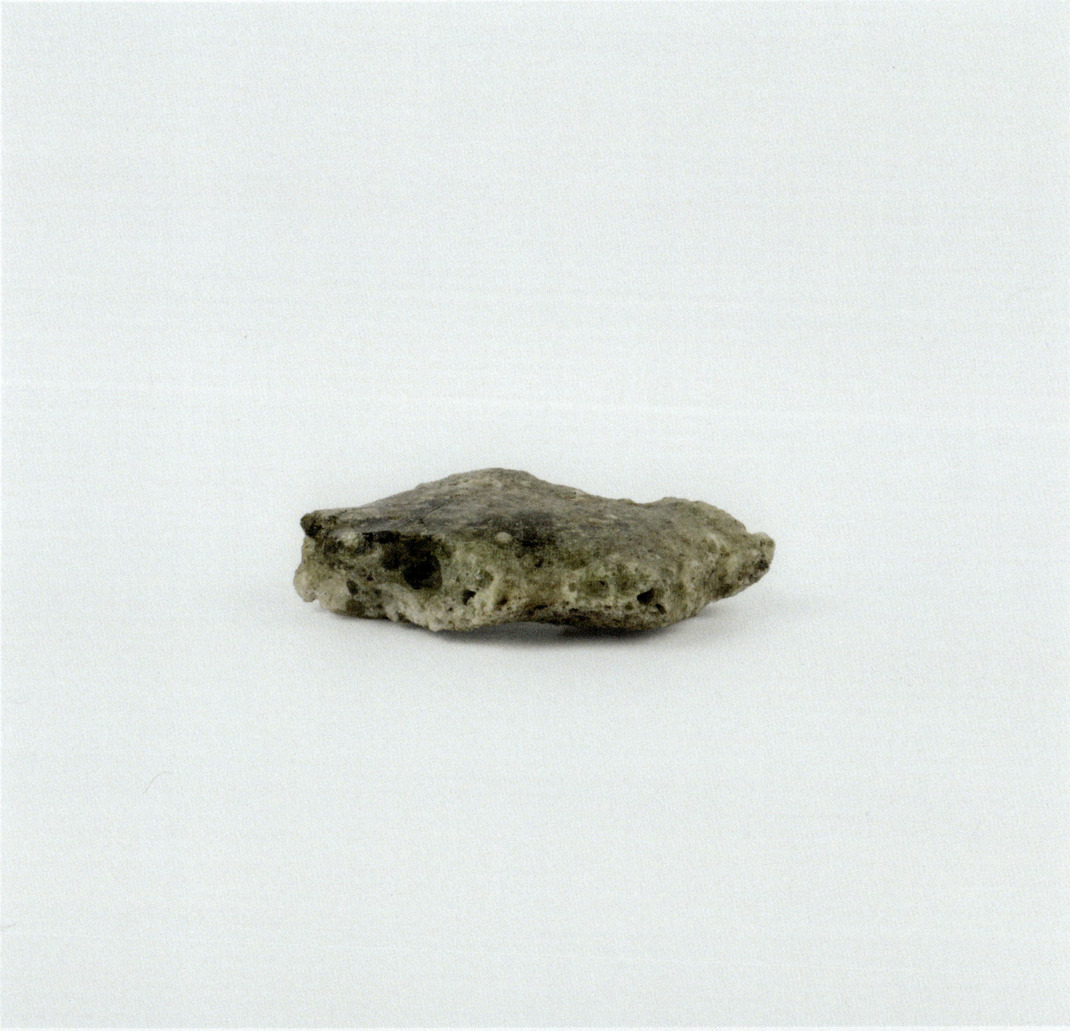

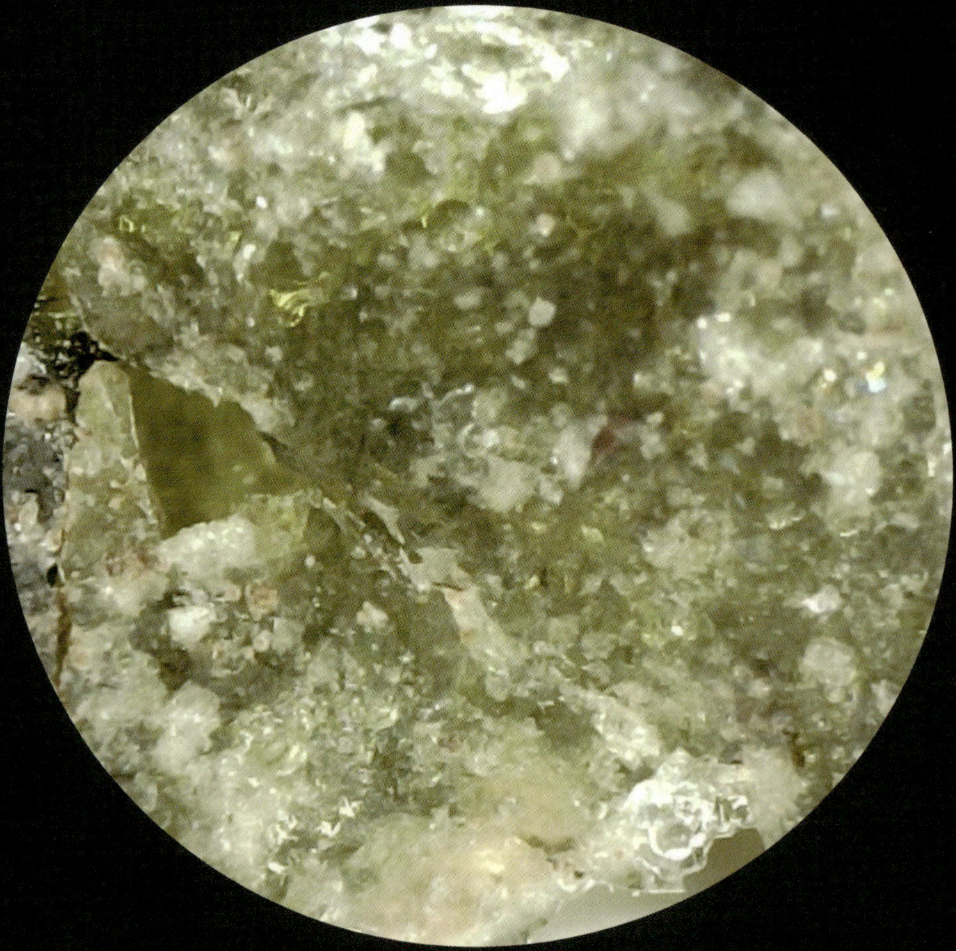

SiO2 · Al2O3 · CaO · Na2O · K2O

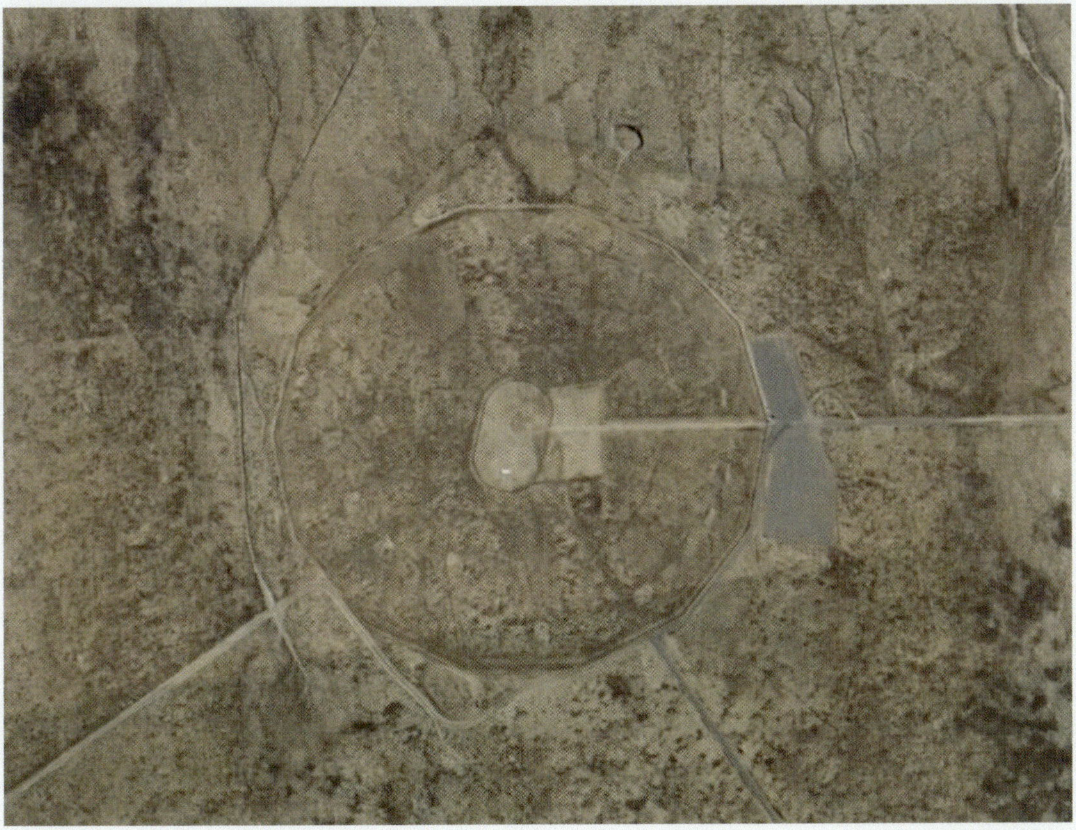

LOCALITY: Trinity Site, New Mexico, United States

COORDINATES 33.6773,-106.4754

List of the 208 anthropogenic mineral species approved by the International Mineralogical Association

ALTERATIONS RECOVERED FROM ORE DUMPS

Bernalite
Boyleite
Delrioite
Gunningite
Hydromagnesite
Kaňkite
Krausite
Krautite
Lansfordite
Metadelrioite
Nesquehonite
Orthoserpierite
Pharmacolite
Ramsbeckite
Rosièresite
Schuetteite
Schulenbergite
Scorodite
Smrkovecite
Wheatleyite
Widgiemoolthalite
Yvonite

ALTERATIONS ASSOCIATED WITH MINE TUNNEL WALLS

Adolfpateraite
Albrechtschraufite
Alwilksinite-(Y)
Andersonite
Apexite
Bayleyite
Běhounekite
Belakovskiite
Bianchite
Bluelizardite
Bobcookite
Calciodelrioite
Canavesite
Cobaltoblödite
Cobaltzippeite
Coquimbite
Dietrichite
Fermiite
Gatewayite
Geschieberite
Goslarite
Gunterite

Hughesite
Jáchymovite
Ježekite
Kegginite
Klaprothite
Kokinosite
Kornelite
Línekite
Magnesiozippeite
Manganoblödite
Marécottite
Mesaite
Mathesiusite
Meisserite
Metamunirite
Morrisonite
Natrozippeite
Nickelzippeite
Oppenheimerite
Ottohahnite
Packratite
Pascoite
Péligotite
Picropharmacolite
Plášilite
Pseudojohannite
Rakovanite
Schindlerite
Swartzite
Sejkoraite-(Y)
Slavkovite
Štěpite
Svornostite
Uranopilite
Vanarsite
Vysokýite
Wernerbaurite
Wetherillite
Zdeněkite
Znucalite
Zýkaite

ALTERATION OF AN EXPOSED ORE BODY

Huemulite

MINERALS ASSOCIATED WITH COAL MINE DUMPS

Koktaite

MINE WATER PRECIPITATES

Alpersite
Bluestreakite
Ferrarisite
Fluckite
Jôkokuite
Jurbanite
Khademite
Kobyashevite
Lannonite
Magnesiopascoite
Martyite
Nickelhexahydrite
Phaunouxite
Phosphorrösslerite
Postite
Rauenthalite
Rossite
Sainfeldite
Wilcoxite

MINERALS FOUND IN SLAG OR SMELTER WALLS

Boleite
Cetineite
Claringbullite
Cumengeite
Cyanochroite
Elyite
Fiedlerite
Georgiadèsite
Glaucocerinite
Kapellasite
Ktenasite
Langite
Laurionite
Lautenthalite
Nealite
Namuwite
Nitrobarite
Paralaurionite
Penfieldite
Posnjakite
Pseudoboleite
Simonkolleite
Serpierite
Thorikosite
Wroewolfeite
Wülfingite
Zlatogorite

MINERALS ASSOCIATED WITH MINE DUMP FIRES

Acetamide
Alum-(Na)

Arsenolite
Bararite
Barberiite
Bazhenovite
Boussingaultite
Butlerite
Cryptohalite
Cuprospinel
Dmisteinbergite
Downeyite
Dypingite
Efremovite
Esseneite
Fluorellestadite
Godovikovite
Guildite
Gwihabaite
Hoelite
Kladnoite
Laphamite
Lausenite
Letovocite
Mascagnite
Mikasaite
Millosevichite
Minium
Mohrite
Ransomite
Ravatite
Rorisite
Rostite
Shannonite
Srebrodolskite
Svyatoslavite
Tinnunculite
Yavapaiite

INTERACTION WITH MINE TIMBERS OR LEAF LITTERS

Devilline
Paceite
Pentahydrite
Hoganite
Metarossite
Monteponite
Nickelboussingaultite
Rabbittite
Rhomboclase
Szomolnokite
Tschermigite
Wupatkiite

MINERALS ASSOCIATED WITH GEOTHERMAL PIPING SYSTEMS

Ammonioborite
Biringuccite
Larderellite
Nasinite
Santite
Sassolite
Sborgite

MINERALS ASSOCIATED WITH ARCHEOLOGICAL ARTIFACTS

Abhurite (tin)
Atacamite (bronze)
Barstowite (lead)
Chalconatronite (bronze)
Cotunnite (lead)
Hydroromarchite (tin)
Phosgenite (lead)
Romarchite (tin)

MINERALS ASSOCIATED WITH THE ALTERATION OF PINE RAILROAD TIE

Arcanite

MINERALS FORMED BY THE ALTERATION OF STORED DRILL CORE

Nickelbischofite

PREHISTORIC SACRIFICIAL BURNING SITES

Fayalite
Forsterite
Stanfieldite
Whitlockite

MISCELANEOUS HUMAN-MEDIATED MINERALS

Calclacite (formed in storage cabinets in museums)
Jedwabite (allegedly from placers, possible hoax)
Niobocarbide (allegedly from placers, possible hoax)
Tantalcarbide (allegedly from placers, possible hoax)

ACKNOWLEDGEMENTS

A special thank you to Brandon Bandy, Rachel Jackson, Andy Bennett, Geneva Skeen, Hannah Kay, Zachary Korol Gold, Hunter Shaw, Nathan Gulick, and Brian Bowman.

This mineral collection was funded in part by Fulcrum Arts and was first published online in 2021 as a contribution to their program *Sequencing,* curated by Geneva Skeen. Additional financial support was granted from the Windward Research & Development Fund.

Designed by Rachel Jackson for Special Effects

Printed at Typecraft in Pasadena, California

ISBN 979-8-9874645-2-6

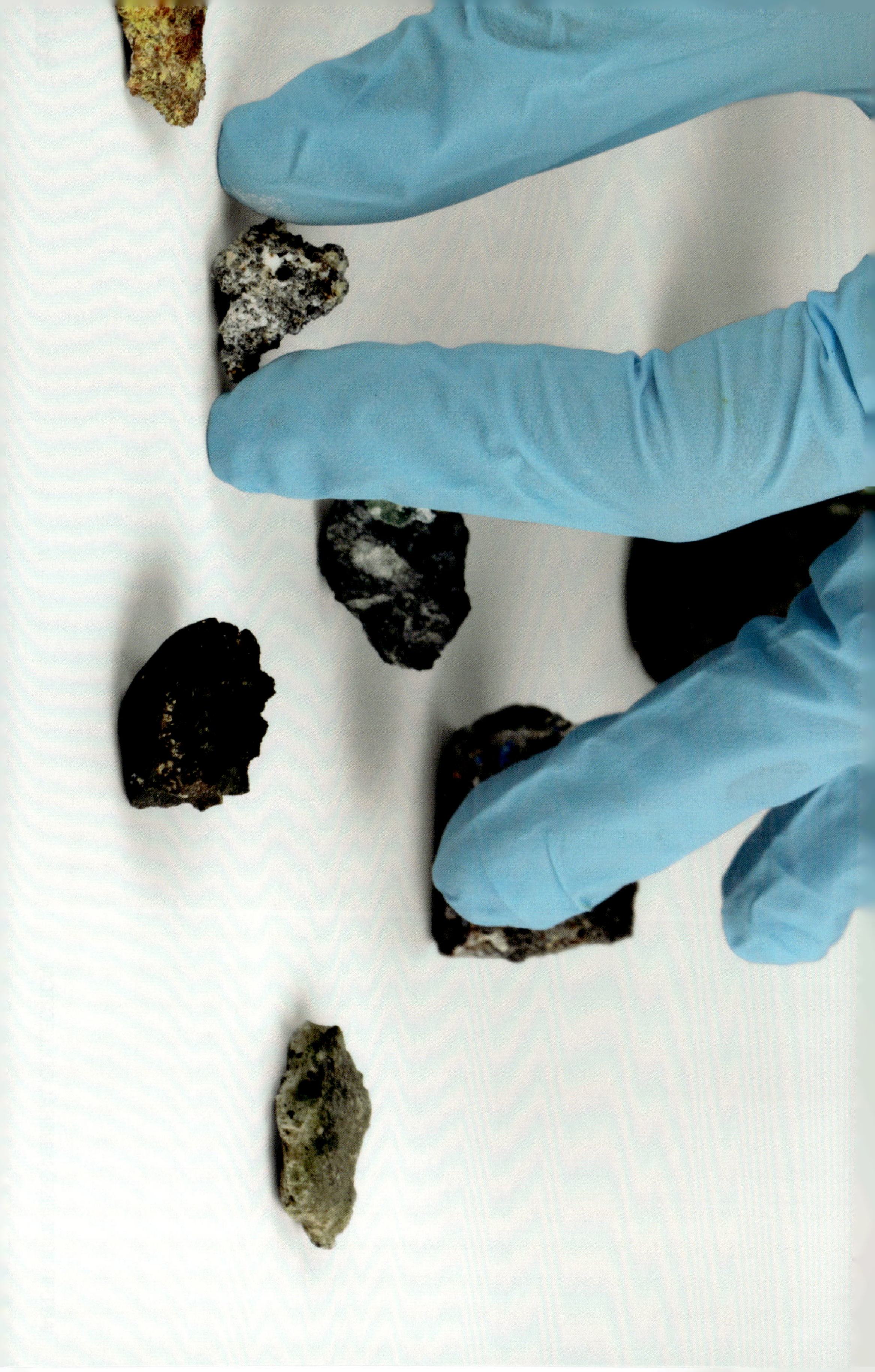

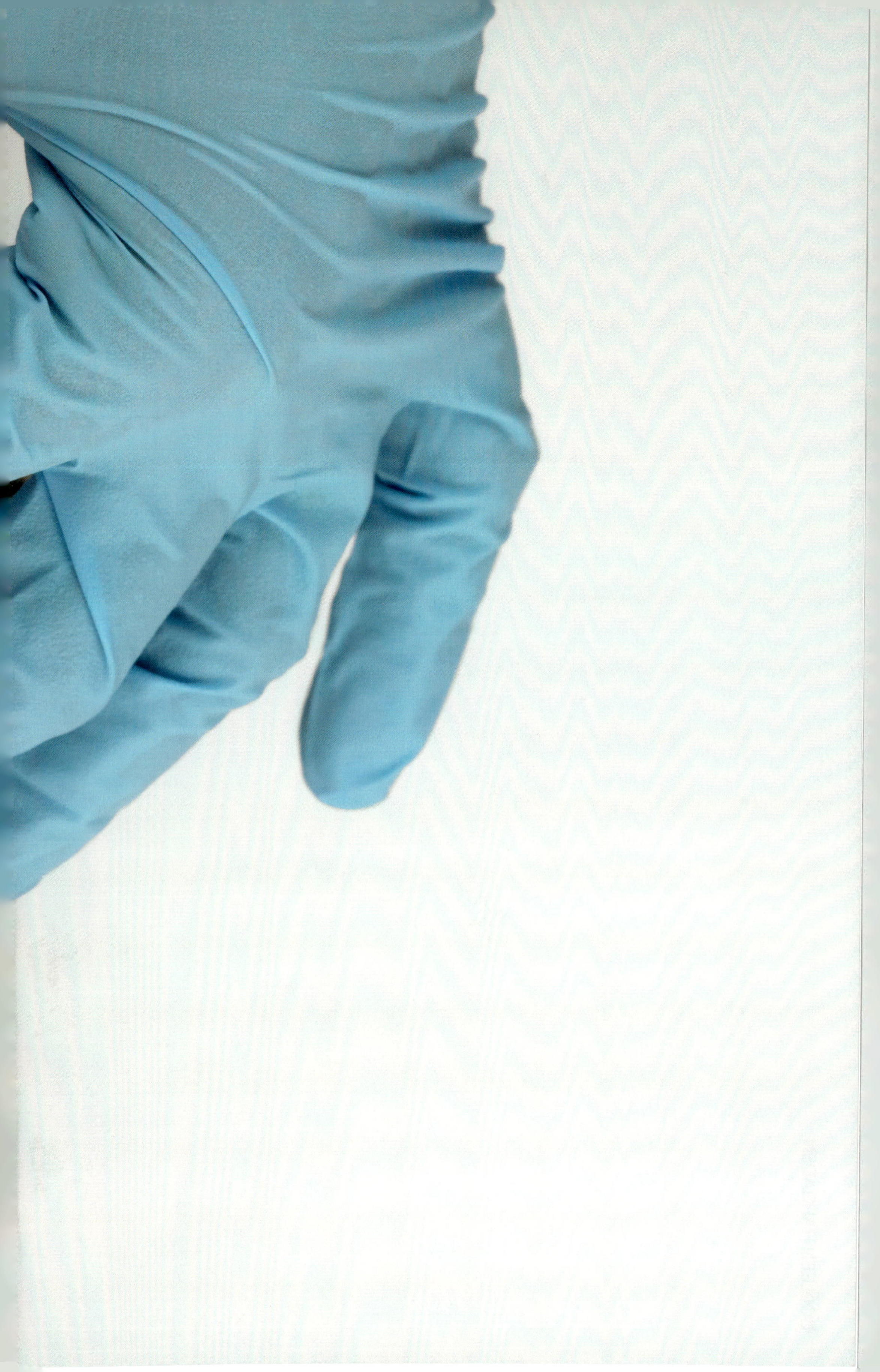